W0082403

Limits to
National Jurisdiction
over the Sea

 VIRGINIA LEGAL STUDIES *are sponsored by the School of Law of the University of Virginia for the publication of meritorious original works, symposia, and reprints in law and related fields. Titles previously published include:*

Central Power in the Australian Commonwealth, by the Rt. Hon. Sir Robert Menzies, former Prime Minister of Australia. 1967.

Administrative Procedure in Government Agencies—Report by Committee Appointed by Attorney General at Request of President to Investigate Need for Procedural Reforms in Administrative Tribunals (1941), reprinted with preface and index 1968.

The Road from Runnymede: Magna Carta and Constitutionalism in America, by A. E. Dick Howard. 1968.

Non-Proliferation Treaty: Framework for Nuclear Arms Control, by Mason Willrich. 1969.

Mass Production Justice and the Constitutional Ideal—Papers and proceedings of a conference on problems associated with the misdemeanor, held in April 1969, under the sponsorship of the School of Law, edited by Charles H. Whitebread, II. 1970.

Education in the Professional Responsibilities of the Lawyer—Proceedings of the 1968 National Conference on Education in the Professional Responsibilities of the Lawyer, edited by Donald T. Weckstein. 1970.

The Valuation of Nationalized Property in International Law—Essays by experts on contemporary practice and suggested approaches, edited by Richard B. Lillich. v. I, 1972; v. II, 1973.

Legislative History: Research for the Interpretation of Laws, by Gwendolyn B. Folsom. 1972.

Criminal Appeals: English Practices and American Reforms, by Daniel J. Meador. 1973.

Humanitarian Intervention and the United Nations—Proceedings of a conference held in March 1972, with appended papers, edited by Richard B. Lillich. 1973.

The United Nations, A Reassessment: Sanctions, Peacekeeping, and Humanitarian Assistance—Proceedings of a conference held in March 1972, edited by John M. Paxman and George T. Boggs. 1973.

Mr. Justice Black and His Books—Catalogue of the Justice's personal library, by Daniel J. Meador. 1974.

Legal Transplants, by Allan Watson. 1974.

The Legal Systems of Africa Series—Volumes published cover Ethiopia (by Kenneth R. Redden, 1968); French-speaking Africa, v. 1 (by Jeswald W. Salacuse, 1969); Congo-Kinshasa (Zaire) (by John Crabb, 1970); Lesotho (by Vernon V. Palmer and Sebastian M. Poulter, 1972); and Somalia Republic (by Haji N. A. Noor Muhammad, 1973).

Limits to National Jurisdiction over the Sea

EDITED BY GEORGE T. YATES III AND
JOHN HARDIN YOUNG

WITH THE COLLABORATION OF
GEORGE D. BILLOCK, JR.

University Press of Virginia
Charlottesville

The publication of this volume is sponsored by the John Bassett Moore Society of International Law

THE UNIVERSITY PRESS OF VIRGINIA
Copyright © 1974 by the Rector and Visitors
of the University of Virginia

First published 1974

Library of Congress Cataloging in Publication Data

Yates, George T
  Limits to national jurisdiction over the sea.
  1. Territorial waters. 2. Continental shelf. I. Young, John Hardin, joint ed.
II. Title.   JX4131.Y37   341.44′8   74-3036   ISBN 0-8139-0572-9

Printed in the United States of America

Foreword

Whatever 1974 is to be called the international "year of," the third United Nations Conference on the Law of the Sea, to be held in Caracas from June 20 to August 29, is already making 1974 for international lawyers the year of the culmination of the response to the disintegration of this body of law. The 1958 and 1960 United Nations Conferences in Geneva produced, as a result of many years' work in the International Law Commission, four basic treaties on the law of the world's oceans and coastal waters. The most notorious failure among the many substantial achievements of the four Conventions in codifying or restating the law of the sea was the inability of the Conferences to agree on a definition of the breadth of each State's territorial sea. Given these Conventions' definition of the high seas as all parts of the sea not included in the territorial sea and the internal waters of a State and the similarly interrelated boundary definitions for such areas as the contiguous zones and the continental shelf, the conference failure has had most far-reaching effects. The most dramatic has been the abandonment of the historic effort, led by the United States, to insist on the three-mile limit and the recent leap seaward, not to six or nine miles, but to what appears to be the general 1974 pre-Conference agreement of a twelve-mile limit. This general agreement has placed even greater emphasis on the necessity, in the view of the principal maritime powers, for the clearest possible definition of the right of innocent passage (*free transit* is the preferred 1974 term of art) through and over straits used for international navigation. Notably, the definition of the continental shelf of the 1958 Geneva Conference is similarly affected by the proposals to fix the territorial sea at twelve miles and so also proposals for newly to-be-defined areas such as the international seabed area and a coastal seabed economic area.

Although the Caracas Conference will be confronting many acute problem areas other than these jurisdiction boundary definitions, such as international marine environment and pollution, fishing and other marine life regulation, the freedom of scientific research, and the creation of treaty institutions for the control and development of the international seabed, it is clear that the sharpest debate and most acute trade-off diplomacy will center on the jurisdictional definitions in light of the possible proposed resolutions of the tension between the needs of the international community and State assertions and claims.

It is therefore most useful to have this volume of essays, all of which are addressed to the central question of the limits to national jurisdiction over the sea. They are a welcome addition and a substantial contribution to the literature, commentary, and proposals for the development of the law of the sea.

Professor Goldie and Mr. Finlay have given us stimulating and well-matched essays on the definition of the continental shelf. One emphasizes the possibilities of continental shelf doctrine being relied upon to move toward international institutions and solutions; the other stresses the risks and concerns of departing from the comforts and relative certainties of the existing national legal regimes. Professor Green and Mr. Butler have carefully set out the several approaches to the jurisdictional and boundary problems by our two giant continental neighbors, the USSR and Canada. Messrs. Ely and Marcoux provide a valuable case study of solutions proposed when applied and tested in the critical area of the South China Sea. The expertise and professionalism of the Office of the Geographer of the Department of State can be appreciated in the personal descriptions by Messrs. Hodgson and McIntyre of particular scientific techniques that are open to States in assessing their national seabed boundary options.

Whether the reader will see the effort to solve the many problems set forth in this volume from the very eye of the summer hurricanes that are surely predicted for the delegations at Caracas, from the safety deep within a governmental or corporate bureaucracy, or from a safe academic haven, he will be most grateful for the assistance of these essays in describing boundaries and setting limits, real or only to be hoped for.

A LBERT H. G ARRETSON

New York, New York
June 1974

Preface

He had bought a large map representing the sea,
Without the least vestige of land:
And the crew were much pleased when they
found it to be
A map they could all understand.

Lewis Carroll
The Hunting of the Snark 15 (1876)

The most vexing and frequently avoided issue in proposals for the establishment of an international regime over the sea, seabed, and appurtenant air space is that of the limits to national jurisdiction. However difficult of resolution the dispute may appear, its settlement must be the principal objective of the new United Nations Conference on the Law of the Sea.[1]

Precedent for avoidance of the issue is clearly evident in the current regime. The approach of the participants at the 1958 Geneva Convention on the Law of the Sea was to skirt the tough issue of national control in order that there might be agreement on related questions. Thus, Article 1 of the Convention on the Territorial Sea and Contiguous Zone[2] declares: "The sovereignty of a State extends, be-

---

1. For the scope of the Conferences' agenda, see G.A. Res. 2750C, 25 U.N. GAOR, Supp. 28, at 26, U.N. Doc. A/8097 (1970), 10 INT'L LEGAL MATERIALS 226 (1971). "By this resolution, the United Nations General Assembly, [d]ecides to convene in 1973 . . . a Conference on the Law of the Sea which would deal with the establishment of an equitable international regime—including an international machinery—for the area and the resources of the sea-bed and the ocean floor, and the subsoil thereof, beyond the limits of national jurisdiction, a precise definition of the area, and a broad range of related issues including those concerning the régimes of the high seas, the continental shelf, the territorial sea (including the question of its breadth and the question of international straits) and contiguous zone, fishing and conservation of the living resources of the high seas (including the question of the preferential rights of coastal States), the preservation of the marine environment (including *inter alia,* the prevention of pollution) and scientific research. . . ." *Id.* at 228. *See also* Stevenson & Oxman, *The Preparations for the Law of the Sea Conference,* 68 AM. J. INT'L L. 1 (1974); Dole & Stang, *Ocean Politics at the United Nations,* 50 ORE. L. REV. 378 (1971); Pollack, *Fisheries Considerations of Ocean Space,* 4 NATURAL RESOURCES LAWYER 676 (1971).
2. Convention on the Territorial Sea and Contiguous Zone, *done* Apr. 29, 1958, [1964] 2 U.S.T. 1606, T.I.A.S. No. 5639, 516 U.N.T.S. 205.

yond its territory and its internal waters, to a belt of sea adjacent to its coast, described as the territorial sea." Although the Convention affords instruction on the measurement of the zone, it glaringly omits the distance to be measured.[3]

This less than satisfying definition of the territorial sea is a key point of reference for the other Conventions. Article 1 of the Convention on the High Seas[4] provides: "The term 'high seas' means all parts of the sea that are not included in the territorial sea or in the internal waters of a State." Article 1 of the Convention on the Continental Shelf[5] leaves uncertain the scope of the continental shelf by utilizing the "definition" of territorial waters in stating that the shelf extends "to the seabed and subsoil of the submarine areas adjacent to the coast but outside the area of the territorial sea, to a depth of 200 metres . . . ." The Continental Shelf Convention further complicates the definition by introducing the exploitability test, which provides for the shelf's boundary to extend "beyond that limit, to where the depth of the superjacent waters admits of the exploitation of the natural resources of the said areas . . . ." Finally, the Convention on Fishing and Conservation of the Living Resources of the High Seas[6] establishes rights and duties with respect to States' activities in the high seas without defining the term "high seas."

In leaving unresolved the issue of the limits of national jurisdiction, the Geneva Conventions perpetuated an atmosphere conducive to competing and ever increasing assertions of national sovereignty. Despite the expanding abilities and ambitions of developed countries to exploit living resources, none of the Conventions forthrightly assured the protection and conservation of these resources outside territorial waters. In response, certain States, notably those of Latin America, following the precedent set by President Truman,[7] unilaterally

---

3. *But cf.* Fitzmaurice, Fisheries Jurisdiction Case (Federal Republic of Germany v. Iceland) (separate opinion), 12 INT'L LEGAL MATERIALS 300, 310 n.1 (1973), where he observes that Article 24 of the Territorial Sea Convention by limiting the contiguous zone to twelve miles from the coastline "implied a territorial sea of *less* than 12 miles extent. . . ."

4. Convention on the High Seas, *done* Apr. 29, 1958, [1962] 2 U.S.T. 2312, T.I.A.S. No. 5200, 450 U.N.T.S. 82.

5. Convention on the Continental Shelf, *done* Apr. 29, 1958, [1964] 1 U.S.T. 471, T.I.A.S. No. 5578, 499 U.N.T.S. 311.

6. Convention on Fishing and Conservation of the Living Resources of the High Seas, *done* Apr. 29, 1958, [1966] 1 U.S.T. 138, T.I.A.S. No. 5969, 559 U.N.T.S. 285.

7. Presidential Proclamation No. 2667, Policy of the United States With Respect to the National Resources of the Subsoil and Seabed of the Continental Shelf, 10 Fed. Reg. 12303 (1945), 13 DEP'T STATE BULL. 485 (1945).

claimed territorial zones as far as 200 miles off their coasts.[8] The controversy between Great Britain and Iceland over cod fishing, escalating to the use of force, illustrates the seriousness of the problem with respect to fishing.[9] Discoveries of mineral deposits in the seabed and the increasing technological feasibility of their extraction further incite national interests and raise the question of the need for an international regime to control and divide the wealth of the seabed. Moreover, the ecological disasters of the *Torry Canyon* and other oil spills as well as the trans-Arctic voyage of the *Manhattan* give rise to widespread concern for the preservation of the ecology and the prevention of international ocean pollution.

The problem does not admit of an easy solution. Even the realization of Carroll's image of a map without land, although it might afford a refreshing change, would not resolve the basic issue of control. The time is rapidly approaching, however, when the issue must be confronted and resolved. To delay is only to foster international disorder.

8. On May 8, 1970, nine Latin American States signed the Montevideo Declaration on Law of the Sea, in which they declared claims to a 200-marine-mile territorial sea and "[t]he right [of littoral States] . . . to delimit their maritime sovereignty and jurisdiction in conformity with their own geographic and geological characteristics and consonant with factors that condition the existence of marine resources and the need for national exploitation; [and] [t]he right to explore, conserve and exploit the living resources of the sea adjacent to their territories . . . [and] . . . the natural resources of the seabed and of the subsoil of the ocean floor out to where the littoral State claims jurisdiction over the sea. . . ." 9 INT'L LEGAL MATERIALS 1081, 1083 (1970). National claims asserted include: Argentina's claim to a 200-mile territorial sea, Decree Law 17,094 (Jan. 4, 1967); Brazil's claim to a 200-mile territorial sea, Decree 1098 (Mar. 25, 1970); Chile's claim to a 200-mile territorial sea, Supreme Resolution No. 179 (Apr. 11, 1953); Costa Rica's claim to a 200-mile fisheries conservation zone, Decree Law No. 739 (Oct. 4, 1949); Ecuador's claim to a 200-mile territorial sea, Executive Accord (Nov. 10, 1966), Decree Law 1542 (Nov. 11, 1966); El Salvador's claim to a 200-mile territorial sea, CONSTITUTION art. 7 (Sept. 14, 1950); Guinea's claim to a 130-mile territorial sea, Decree No. 224 (June 3, 1954); Nicaragua's claim to a 200-mile exclusive fishing zone, Executive Decree 1-L (Apr. 5, 1965); Panama's claim to a territorial sea of 200 miles, Law No. 31 (Feb. 3, 1967); Peru's claim to an exclusive fishing jurisdiction of 200 miles, Executive Decree of Aug. 1, 1947; Uruguay's claim to a 200-mile territorial sea, Decree of May 12, 1969, Decree of Dec. 13, 1969. *See generally* Lecuona, *The Equador Fisheries Dispute,* 2 J. MARITIME LAW & COMMERCE 91 (1970); Loring, *The United States-Peruvian Fisheries Dispute,* 23 STAN. L. REV. 391 (1971).

9. *See* Fisheries Jurisdiction Case (United Kingdom v. Iceland), [1972] I.C.J. 12, *reprinted in* 11 INT'L LEGAL MATERIALS 1069 (1972) (order concerning the request for indication of interim measure of protection); Fisheries Jurisdiction Case (United Kingdom v. Iceland), [1972] I.C.J. 181, *reprinted in* 11 INT'L LEGAL MATERIALS 1077 (1972) (order concerning the question of the Court's jurisdiction); Fisheries Jurisdiction Case (United Kingdom v. Iceland), 12 INT'L LEGAL MATERIALS 290 (1973) (judgment on the question of jurisdiction).

This collection of esays is not offered to espouse the ultimate solution to the question of national jurisdiction over the sea. Rather, the writers focus on two areas of the controversy: Jurisdiction over the Continental Shelf and the Seabed, and the national response: Contemporary State Practice. It is hoped that these essays will further recognition of the need for an international regime of the seas offering specific provisions to limit national jurisdiction. The issue can no longer be avoided.

We wish to thank the contributors to the volume, who have produced writings of considerable merit. We wish also to acknowledge the kind permission of the *New York Law Forum* to draw upon material published and copyrighted by them for part of the material appearing in chapter 1. Books of this kind would not be possible without the help and encouragement of such professors as Richard B. Lillich, who guided us through the process of publication; Carl McFarland, who gave us valuable insights into the law of the sea; and Monrad Paulsen, who served as our Dean and friend in securing support for publication of this volume. We further wish to acknowledge the editorial and administrative assistance of George Billock, Vice President of the John Bassett Moore Society of International Law; we here thank Michael Weaver and William Seitz, who checked citations in two of the Chapters, Mrs. Oleta J. Hamilton for the tedious task of preparing the index and Virginia Holden for her support. Finally, we record our appreciation to the Publications Committee of the University of Virginia School of Law and the University Press of Virginia for their advice and support and to the Doherty Foundation, which most graciously funded the book.

GEORGE T. YATES III
and
JOHN HARDIN YOUNG

Charlottesville, Virginia
June 1974

Contents

Maps

Tables

Figure

Jurisdiction over
the Continental Shelf

I

Delimiting Continental Shelf Boundaries

L. F. E. Goldie

Introduction

Determining the basis for the delimitation of the continental shelf creates problems of demarcating its upper and outer limits and the boundaries between the legal shelves of adjacent and opposite States abutting on a common geographical shelf. The introduction of the concept of a coastal State's adjacent continental shelf as the "natural prolongation of its land territory into and under the sea"[1] as a criterion for determining the boundaries of States' submarine areas on common shelf areas adds further dimensions to the debate. Islands and islets further complicate the delimitation process. Interestingly, these problems have received little attention in the scholarly and professional literature.[2] The delimitation of the shelf also raises problems of the scope of the "exploitability test" as set forth in Article 1 of the Continental Shelf Convention[3] and of proposals for an "intermediate zone" or a "trusteeship area."[4]

Apart from semantic difficulties in setting precise limits to coastal States' adjacent continental shelves, political drives, varying from economic need to cartographical chauvinism and a desire "to keep up with the Joneses,"[5] have complicated and displaced the analytical problems of determining the scope and limits of the continental shelf doctrine. This phenomenon can be illustrated by the development of recent doctrines or, rather, verbalizations of claims that are currently being foisted on the maritime community of nations in the guise of

1. *See* [1969] I.C.J. 3, 22. *See also* note 14 *infra* (for additional citations to the use of this term by the Court) and accompanying text; pp. 7–13 and 39–44 *infra*.
2. An exception to this general oversight in the literature of maritime and continental shelf boundaries is to be found in the pioneering study Ely, *Seabed Boundaries Between Coastal States: The Effect to be Given Islets as "Special Circumstances,"* 6 INT'L LAWYER 219 (1972).
3. Convention on the Continental Shelf, *done* Apr. 29, 1958, [1964] 1 U.S.T. 471, T.I.A.S. No. 5578, 499 U.N.T.S. 311, (effective June 10, 1964) [hereinafter cited as Continental Shelf Convention].
4. These proposals are discussed at pp. 63–74 *infra*.
5. *See* Boggs, *National Claims to Adjacent Seas,* 41 GEOGRAPHICAL REV. 185 (1951).

legal principles. These imaginative twentieth-century echoes from the Dark Ages' *droit de bris* have been given such names as the "economic," "patrimonial," and "closed" seas.[6] Analytically, they tend to be meaningless emotive terms evocative of demands. Politically, they verbalize assertions of power, so that their significance and content are variables governed by what can feasibly be demanded at any given time. As with those doctrines, so with much of the contemporary controversy about the outer limits of the continental shelf; investigation should be clarified by separating the problems that can be the subject of analytical appraisal from those that are condemned essentially to political explanations—that is, explanations in terms of power rather than reasoned analysis.

If juridically viable solutions are to prevail over emotive sloganeering in bringing about a new equitable and impartial distribution of rights among individual States, regional and other interest groups, and the global community, they should be concretely formulated in conventions admitting of legal analysis. These conventions, in their turn, should be susceptible of elucidation by reasoned investigation, classification under some established category, and subsumption under an objective value judgment whose content is known and accepted. They should, furthermore, provide for the settlement of disputes, when good-faith bargaining[7] and other informal means of conciliatory settlement and arbitration fail, by invoking the compulsory jurisdiction of the International Court of Justice. This procedure should be conferred either by the conventions themselves or by an accompanying Protocol similar, for example, to the one accompanying the four 1958 Geneva Law of the Sea Conventions.

The Continental Shelf's Upper Limits

Article 3 of the Continental Shelf Convention carefully separates the regime of the continental shelf from that of the free high seas (including the superjacent air column). It provides: "The rights of the coastal State over the continental shelf do not affect the legal status of

---

6. For a critical review of these maritime claims, *see* Goldie, *International Law of the Sea—A Review of States' Offshore Claims and Competences,* 24 NAVAL WAR COLLEGE REV., Feb. 1972, at 43, 49–51.

7. For a discussion of this emerging international law concept governing disputes over continental shelf delimitations, *see* pp. 13–28. *See also* Goldie, *The North Sea Continental Shelf Cases—A Ray of Hope for the International Court?,* 16 N.Y. L.F. 325, 359–67 (1970). This writer's reuse of materials and analysis first published in that Article is acknowledged.

the superjacent waters as high seas, or that of the air space above those waters." The first boundary to be discussed, then, is that of the continental shelf and the free high seas above the shelf, beyond the limits of national jurisdiction.

In his pioneering essay on the continental shelf, Sir Cecil Hurst described his proposed severance of the volume of the high seas above the continental shelf from the seabed and subsoil as follows:

Hitherto, it has, I believe, been generally assumed that the limits of a State's sovereignty is a vertical straight line stretching upwards and downwards *ad infinitum* from the starting point. Was the Continental Shelf policy intended to introduce a new system? A system under which the limits of a State's sovereignty would be a line which made a gigantic zig-zag. A line, that is to say, which, starting from the bed of the sea at the limit of territorial waters went upward vertically to the surface of the waters and on upwards *ad infinitum* through the airspace—but instead of going downwards *ad infinitum* went laterally in a more or less horizontal direction along the seabed until it reached the limits of the Continental Shelf and there made an angle and went down *ad infinitum*.[8]

The flat horizontal boundary that Hurst envisaged lay on the surface of the seabed, including that surface in the regime of the continental shelf (and not that of the high seas) where the shelf extended beyond the limits of national jurisdiction. The above quotation also makes explicit the thesis underlying the British instruments claiming offshore submarine areas, other than those defining and regulating the continental shelf rights of the United Kingdom.[9] These instruments all assumed the shelf to be territory and assumed the competence of the coastal State to be plenary sovereignty over the seabed and its subsoil (although they were careful not to include any claim to the waters of the free high seas and air above the shelves as brought within British sovereignty). In fact, their formulation was based upon proclamations made under the Colonial Boundaries Act of 1895.[10] They were constitutive and annexatory by virtue of the formula that proclaimed "boundaries" to be "extended" in the three-dimensional fashion Hurst prescribed.

While embracing the thesis that the water column should be separated from the continental shelf, the position of Admiral M. W. Mouton should be distinguished from Hurst's. He argued that the

8. Hurst, *The Continental Shelf*, in INTERNATIONAL LAW, THE COLLECTED PAPERS OF SIR CECIL HURST 152, 162 (1950).
9. *See* long title and §1 (1) of the Continental Shelf Act 1964, c. 29. This legislation was in terms of the "sovereign rights" accorded under Article 2 of the Continental Shelf Convention.
10. 58 & 59 Vict., c. 34.

seabed should be "divorced" from the continental shelf and "married" to the high seas.[11] By this, Mouton meant that the seabed should not be included within the terms of reference of the continental shelf doctrine but should be brought under the regime of the high seas. He further considered that the rights of the coastal State should be restricted to the rights of usufruct or, at most, of "ownership" of only the minerals contained in the shelf's subsoil. Like Hurst, however, Mouton saw the existing continental shelf doctrine as purporting to give territorial rights and competences to coastal States. His advocacy was to resist this trend. In so arguing, he was, in fact, attacking a straw man. Neither the continental shelf doctrine nor the Convention accords plenary territorial authority to coastal States, but only specific competences limited to fulfilling particular purposes. From the point of view of the purposive and limited qualities of coastal States' competences with respect to the continental shelf, the divorce and the wedding become irrelevant, since the surface of the seabed, whether shelf or high seas, can still be the scene of exploring the shelf ancillary to the exploitation of its resources. The notion of purposive specific competences, moreover, has advantages similar to those for which Mouton argued in his proposal that coastal States should merely "own" the minerals in the shelf's subsoil. On the other hand, it has the advantages over Mouton's concept of dispensing with metaphysical problems conjured up by thoughts of States owning minerals that are unknown to the owners—and possibly unconceived of by them—and of providing a functional approach that looks to conduct rather than to the reification of concepts.

Adjacent and Opposite States

Distributive Equity

In the *North Sea Continental Shelf Cases,*[12] the Federal Republic of Germany argued that the apportionment of the North Sea continental shelf should be in terms of distributive justice whereby each State received "a just and equitable share" of the divisible area.[13] The Court, however, held this thesis to be

11. M. MOUTON, THE CONTINENTAL SHELF 283–86 (1952).
12. [1969] I.C.J. 3.
13. Reply of Federal Republic of Germany, 1 North Sea Continental Shelf Cases, I.C.J. Pleadings 389, 423 (1968) [hereinafter cited as 1 Pleadings]. Germany contended: "[A]n equitable apportionment of the continental shelf of the North Sea among the surrounding States could not be achieved by determining the boundary

wholly at variance with what the Court entertains no doubt is the most fundamental of all the rules of law relating to the continental shelf, enshrined in
Article 2 of the 1958 Geneva Convention, though quite independent of it—
namely that the rights of the coastal State in respect of the area of continental
shelf that constitutes a natural prolongation of its land territory into and
under the sea exist *ipso facto* and *ab initio,* by virtue of its sovereignty over the
land, and as an extension of it in an exercise of sovereign rights for the purpose
of exploring the seabed and exploiting its natural resources.[14]

Thus, the Court reviewed the totality of the continental shelf of
each North Sea State as already appurtenant to that State. Only the
problem of demarcating the boundaries between the separate shelves
remained. The Court also rejected any idea of the North Sea continental shelf as being something held in common to be divided
equitably among the coastal States. This rejection applied, *a fortiori,*
to the shelf's southeastern sector (bounded by the equidistance lines
between Denmark and Norway and between the United Kingdom and
the Continent). The significance of the term *delimitation* in this
context was pointed out by Judge Morelli in his dissenting opinion,
where he neatly illustrated the distinction between dividing a common
property in equitable shares and determining the boundaries between
the coastal States' appurtenant continental shelves.[15]

---

lines between each pair of adjacent or opposite States as an isolated act. The
boundary problem must rather be considered as a joint concern of all North Sea
States, taking into account the effect of each boundary on the apportionment as a
whole." *Id. See also id.* at 76.

Be that as it may, the Federal Republic's view of distributive justice and equity
was encapsulated in the following thesis:

"(I) In apportioning the continental shelf among the coastal States, the
breadth of their coastal frontage facing the North Sea should be the principal criterion for evaluating whether the area allocated to one of these States is a just
and equitable share.

"(II) The most equitable apportionment of the continental shelf among the
coastal States would be a sectoral division based on the breadth of their coastal
frontage facing the North Sea.

"(III) As to the delimitation of the continental shelf between the Parties, the
equidistance method cannot find application, since it would not apportion a just
and equitable share to the Federal Republic of Germany.

"(IV) The boundary line dividing the continental shelf between the Parties
must be settled by agreement in accordance with the judgment of the Court." *Id.*
at 89.

14. [1969] I.C.J. at 22. The majority stressed the notion of the continental shelf as the
"natural prolongation of the coastal State's land territory" throughout the
Judgment, see *id.* at 29, 30, 31, 36, 37, 39, 40, 43, 44, 47, 51.

15. Judge Morelli stated: "Without a doubt a situation can exist which gives rise to a
problem of *delimitation,* namely the problem of ascertaining how a certain area of

Natural Prolongation and the Question of the Juristic
Inevitability of the Equidistance Principle

Rather than leave the general problem of delimitation or demarcation
to a theory of general distributive equity, the Court justified its thesis
that the parties were under a duty to negotiate the boundaries of their
existing and separate appurtenant continental shelves by resorting to
a slogan that has, seemingly, an avant-garde ring. The majority said
that, in connection with the continental shelf, as in territorial waters
and contiguous zone matters,"the land dominates the sea."[16] And the
"land" in this case was clearly intended to be the land of each coastal
State. It thus followed that the Court would apply a system of demarcation in determining which seabed areas are natural prolongations of
which States. It might seem that Article 6, paragraph 2, of the Continental Shelf Convention would supply a ready answer, since it appears to have a character of "juristic inevitability"[17] for the application of the "natural prolongation" theory. This notion of juristic
inevitability was well defined by Judge Tanaka when he said:

The delimitation itself is a logical consequence of the concept of the continental shelf that coastal States exercise sovereign rights over their own continental shelves. Next, the equidistance principle constitutes the method which
is the result of the principle of proximity or natural continuation of land territory, which is inseparable from the concept of continental shelf. Delimitation
itself and delimitation by the equidistance principle serve to realize the aims
and purposes of the continental shelf as a legal institution. . . .

. . . .

The method of logical and teleological interpretation can be applied in the
case of customary law as in the case of written law. Even if the Federal Republic recognizes the customary law character of only the fundamental concept incorporated in Articles 1-3 of the Convention, and denies it in respect of
other matters, she cannot escape from the application of what is derived as a
logical conclusion from the fundamental concept,—a conclusion which, in
respect of the delimitation of the continental shelf, would reach the same result as Article 6, paragraph 2, of the Convention.[18]

---

the continental shelf is *already apportioned* among two or more States. This
operation of delimitation has nothing to do with the *sharing out,* among two or
more States, of something common to those States." *Id.* at 199.

16. *Id.* at 51. For a similar sentiment, differently formulated, however, see *id.* at 29.
17. *Id.* at 199.
18. *Id.* at 181. *See also* the dissenting opinion of Judge Morelli, *id.* at 202, where he
wrote: "I consider the rule of general international law prescribing the equidistance
criterion for the delimitation of the continental shelves of various Staes to be a
necessary consequence of the apportionment effected by general international law
on the basis of contiguity. I am therefore of the opinion that it is not necessary to

The majority rejected this thesis[19] with a firmness equal to its rejection of the argument asserting that the shelf should be divided among the coastal States in terms of distributive justice.[20] They were persuaded that customary international law required the application of the equidistance rule in this case.[21]

Although it rejected the equidistance rule as a rule governing the lateral delimitation of the adjacent continental shelves of the three littoral States involved in the dispute, the Court decided that the situation was not, as the Federal Republic argued, to be equitably resolved by distributive principles of apportionment,[22] nor was it for the "unfettered appreciation of the Parties."[23] Rather, they were under an obligation to negotiate the requisite delimitations "with a view to arriving at an agreement, and not merely to go through a formal process of negotiations as a sort of prior condition" for the application of the equidistance rule.[24]

Judge Bustamente y Rivero agreed with the Court that the parties were under an obligation to enter into "meaningful" negotiations in which equitable principles of demarcation should be applied.[25] But in his separate opinion,[26] he considered that the majority's premise that

---

ascertain if a specific custom has come into existence in this connection. State practice in this field is relevant not as a constitutive element of a custom which creates a rule, but rather as a confirmation of such rule."

19. *Id.* at 28. Although dissenting from the majority Judgment, Judge Sorensen agreed with this aspect of its evaluation of Article 6, paragraph 2. He disagreed with the thesis that Article 6, paragraph 2 was opposable to the Federal Republic on the ground that it had become a rule of customary international law that the Federal Republic had not consistently refused to recognize "as an expression of generally accepted rules of international law." *See id.* at 247–49.

20. The Court said: "More important is the fact that the doctrine of the just and equitable share appears to be wholly at variance with what the Court entertains no doubt is the most fundamental of all the rules of law relating to the continental shelf, enshrined in Article 2 of the 1958 Geneva Convention, though quite independent of it,—namely that the rights of the coastal State in respect of the area of continental shelf that constitutes a natural prolongation of its land territory into and under the sea exist *ipso facto* and *ab initio,* by virtue of its sovereignty over the land, and as an extension of it in an exercise of sovereign rights for the purpose of exploring the seabed and exploiting its natural resources." *Id.* at 22.

21. On the majority's rejection of the argument proposing the customary law underpinnings of the equidistance rule, see *id.* at 38–45. *See also id.* at 67 (separate opinion of Judge Jessup); *id.* at 85–87 (separate opinion of Judge Padillo Nervo).

22. *Id.* at 22–23, 32.

23. *Id.* at 46.

24. *Id.* at 47. *See also id.* at 53.

25. *Id.* at 65. *See also id.* at 58–59.

26. *Id.* at 57–65.

a coastal State's continental shelf constituted a "natural prolongation of its land territory" called for further guidelines.[27] He said that the "principle of convergence" was "normal for the delimitation of the shelves in this kind of sea."[28] He argued that this principle was a corollary of the natural prolongation theory under the circumstances of the geographical configuration of the North Sea. Judge Bustamente y Rivero contended:

In this kind of configuration, the natural prolongation of the territory of each State, starting from the shore, moves in a seaward direction towards the central area of the sea under consideration; while the lateral boundary lines of each shelf naturally and necessarily converge toward the same central area. The principle of convergence is therefore normal for the delimitation of the shelves in this kind of sea unless the Parties agree upon another solution.[29]

The purpose of this lateral delimitation of boundaries as regards the principle of convergence is to avoid the overlapping of the natural prolongations of the States' land territories. Furthermore, it has the effect of shaping each State's appurtenant continental shelf approximately in the form of a trapezium or triangle, "according to whether the central maritime area is more or less elongated or, on the contrary, more nearly circular."[30] He then laid down the two following specific guidelines as applicable to the German-Danish and German-Dutch delimitations:

(i) the delimitation will be made only beyond the partial boundary lines determined by the treaties of 1 December 1964 and 9 June 1965 already cited (points D and B on [Map 2]);

(ii) the extremities of the two lateral boundary lines to be drawn will meet the line or, as the case may be, the point indicating the western side or apex of the German shelf . . . . It is for the Parties to choose the method or methods for carrying out this lateral delimitation in conformity with the terms of the Special Agreements now in force, as well as to combine those methods with the principle of equity, as contemplated in paragraph 85 of the Judgment.[31]

The question of determining the apex (west side) of the Federal Republic's shelf was to be a matter for the parties to settle between themselves. In such negotiations, however, the Agreement of March

27. *Id.* at 22. *See also id.* at 29, 30, 31, 36, 37, 47, 51. This concept was also mentioned by Judge Bustamente y Rivero, *id.* at 58, 61; Judge Ammoun, *id.* at 116; and Judge Tanaka, *id.* at 180.
28. *Id.* at 61. *See also* Memorial of the Federal Republic of Germany, 1 Pleadings 79–84, 85 (Fig. 21).
29. [1969] I.C.J. at 61.
30. *Id.*
31. *Id.* at 62.

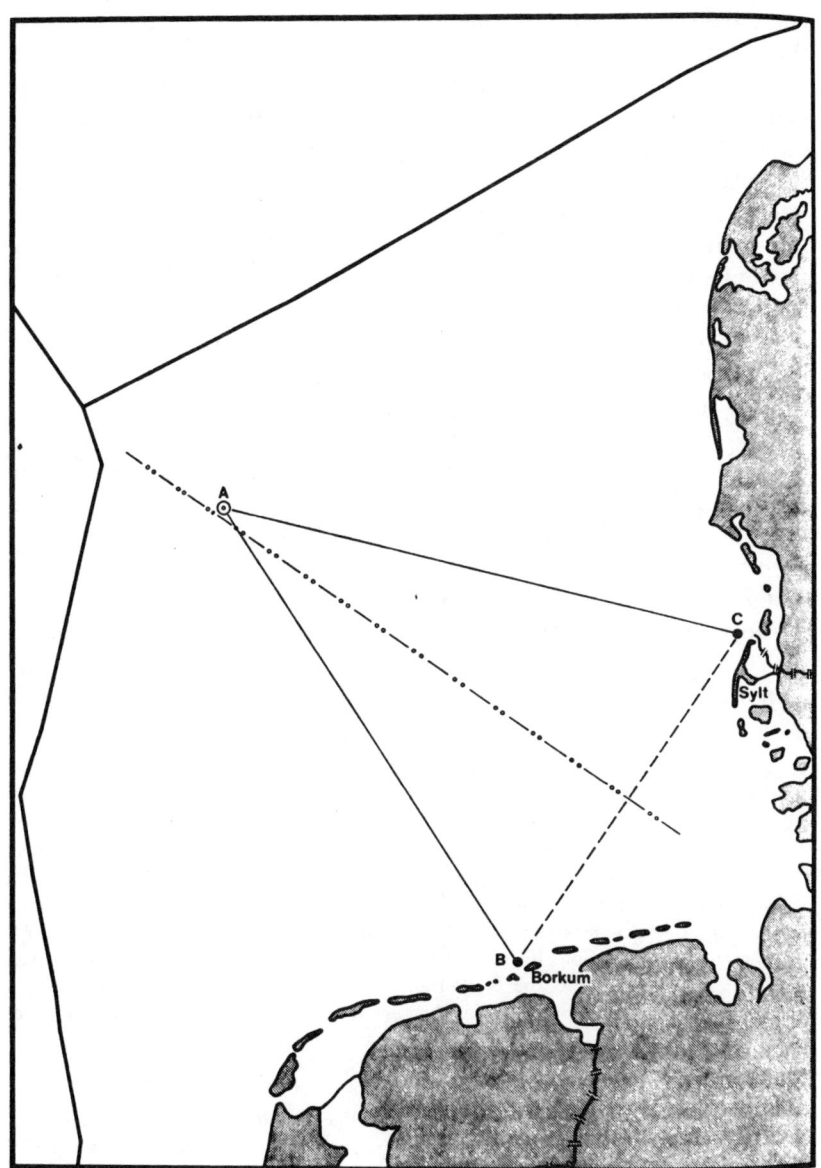

1. Equidistance lines for Federal Republic of Germany and Denmark
([1969] I.C.J. 3, at 153)

31, 1966, between the Netherlands and Denmark was not opposable to the Federal Republic.[32] (By this treaty Denmark and the Netherlands had agreed to allocate continental shelf areas to each other, applying the equidistance principle in such a manner as to preclude the area left over as the Federal Republic's shelf from reaching a median position in the North Sea). Judge Bustamente y Rivero reinforced his argument that the parties were obliged to include the delimitation he proposed in their negotiated settlement by noting that the apex of the lateral boundary lines was at the median line of the shelf. This line, he found, had become consecrated in a "regional customary law" as a result of its use "by the majority of the coastal States in the agreements for delimitation of their shelves."[33]

Judge Ammoun, on the other hand, considered the "equidistance–special circumstances" rule to be applicable. But the "special circumstances" of the southeast section of the North Sea, in his view, required that "equidistance" take account of the concave physical configuration of the portion of the North Sea coast in dispute. The obligatory character of this mode of demarcation flowed from "a general principle of law," namely, equity.[34] He proposed that the appropriate demarcations should subtend a base line drawn between a first point offshore from the island of Borkum and a second point offshore from the island of Sylt. The German share of the North Sea's shelf would then be defined as within two "equidistance lines" (which do not appear, in fact, to be equidistance lines on his map)[35] extending

32. His Honor agreed that two other relevant agreements—those between the United Kingdom and the Netherlands, [1967] Gr. Brit. T.S. No. 23 (CMND. 3253), and the United Kingdom and Denmark, [1967] Gr. Brit. T.S. No. 35 (CMND. 3278)—were not within the Court's jurisdiction. On the other hand, and "[f]rom the hypothetical point of view," he envisaged the possibility of an Anglo-German Settlement "in which the Netherlands and Denmark would acquiesce," establishing a "small section of Anglo-German median line, or simply a point . . . ." [1969] I.C.J. at 64. Alternative suggestions included a tripartite agreement between the Federal Republic, Denmark, and the Netherlands fixing the mathematical position of a "German-British median line" for the purpose of meeting the lateral boundaries of those countries' shelves.

33. *Id.* at 62.

34. *Id.* at 150.

35. Judge Ammoun, *id.* at 152, defined this base line as follows: "In the present case, it is by taking as the starting-point the intersection of the straight baselines marking the coastal fronts of the Federal Republic and Denmark, with due regard for the partial delimitation agreed upon, that the equidistance line between the respective continental shelf areas of those two States could be fixed; and it is by taking as the starting point the intersection of the said baseline of the Federal Republic and that of the Netherlands, that the equidistance line between the two latter States, again with due regard to the agreed partial delimitation, could be fixed. This would be

to an apex beyond that allowed by the equidistance lines proposed by Denmark and the Netherlands, but not extending to the median line, that is, not as far to the west as the lines provided by Judge Bustamente y Rivero's "principle of convergence."[36]

An International Law Obligation to Bargain in Good Faith?

The Source of the Duty to Negotiate the Shelf's Delimitation

The Court's most novel contribution to international law in the *North Sea Continental Shelf Cases* was its decision that "the parties are under an obligation to enter into negotiations with a view to arriving at an agreement, and not merely to go through a formal process of negotiation."[37] Clearly, this duty was seen to be as incumbent on the Federal Republic of Germany as on the Netherlands and Denmark. Accordingly, since Article 6 was held to be not opposable to the Federal Republic, this obligation to negotiate would not appear logically to stem from the references to settlement by agreement formulated in that Article.[38] The Court identified the provenance of its requirements of "delimitation by mutual agreement and delimitation in

---

done in two separate operations. The area appertaining to the Federal Republic would be contained between the two equidistance lines and would extend out to sea as far as their point of intersection."

36. *Id.* at 61. Judge Ammoun does not give a general written description of his demarcation that could be plotted on a map, nor does he give a definition in terms of courses and distances. Accordingly, it is necessary to consult the map, *id.* at 153, to see what His Honor intended. *See* Map 1.

37. *Id.* at 47.

38. Clearly, the Court could not consider the requirement in Article 6, paragraphs 1 and 2, to the effect that the continental shelf boundaries of opposite and adjacent States "shall be determined by agreement among them" as embodying the customary international law of the shelf, without undermining its whole argument that the equidistance–special circumstances rule was not opposable to the Federal Republic. One might infer from its silence on the subject, however, that the majority saw the relation between the general law duty to negotiate and Article 6 as being parallel to the relation of the general international law "obligation not to impede the laying or maintenance of submarine cables or pipelines on the continental shelf seabed" with Article 4, or to that of "the general obligation not unjustifiably to interfere with freedom of navigation, fishing and so on" with Article 5, paragraphs 1 and 6. The Court said that these rights and duties "all relate to or are consequential upon principles or rules of general maritime law, very considerably ante-dating the Convention," and added that they "were mentioned in the Convention, not in order to declare or confirm their existence, which was not necessary, but simply to ensure that they were not prejudiced by the exercise of continental shelf rights as provided for in the Convention." *Id.* at 39.

accordance with equitable principles" as the Truman Proclamation. It
added that the obligation to delimit boundaries in these two conjoint
ways has "underlain all the subsequent history of the subject."[39]

If the duty to negotiate continental shelf boundaries is, on the one
hand, binding on the Federal Republic but, on the other, not a customary rule of the continental shelf (since it does not occur in the first
three articles of the Convention), then there is a question of the
Court's authority to impose this duty. First, it should be noted that
the Court did reject an equation of the duty to negotiate with a
freedom to delimit boundaries weighted merely by the bargaining
assets, strategies, premises, and powers the parties could bring to the
negotiating table. Second, it expressly asserted that the situation was
not "one for the unfettered appreciation of the Parties," and there
were, therefore, still rules and principles of law to be applied."[40] The
applicable rules (that the Court saw as equitable) were:

(a) the parties are under an obligation to enter into negotiations with a view to
arriving at an agreement, and not merely to go through a formal process of
negotiation as a sort of prior condition for the automatic application of a
certain method of delimitation in the absence of agreement; they are under an
obligation so to conduct themselves that the negotiations are meaningful,
which will not be the case when either of them insists upon its own position
without contemplating any modification of it;

(b) the parties are under an obligation to act in such a way that, in the particular case, and taking all the circumstances into account, equitable principles are applied,—for this purpose the equidistance method can be used, but
other methods exist and may be employed, alone or in combination, according
to the areas involved;

(c) for the reasons given in paragraphs 43 and 44, the continental shelf of any
State must be the natural prolongation of its land territory and must not en-
croach upon what is the natural prolongation of the territory of another
State.[41]

Clearly, these rules are intended to spell out the equity the Court
saw as necessarily joined with the duty to negotiate in continental shelf
practice ever since the Truman Proclamation. Furthermore, when
held within the confines of the rules, negotiation provides the
necessary means for determining the delimitations of each coastal
State's natural prolongation seaward of its land territories. This combination of negotiation-plus-equity thus stands in place of Judge

39. *Id.* at 33.
40. *Id.* at 46.
41. *Id.* at 47.

Tanaka's and Judge Morelli's view of the equidistance rule[42]—the necessary means of knowing where each State's natural prolongation bordered those of the others.

If Article 6 did not restate the customary international law of the continental shelf (and the majority had already held that it did not), the Court would have been contradicting itself if the effect of its reference to the provenance of the duty to negotiate was to give a customary law authority to the call for settlement by agreement in that Article. Hence, it is necessary to look outside the Continental Shelf Convention to find the authority for the Court's decision. The submission here is that, although the Court did not invoke Article 33 of the United Nations Charter, that provision—since it obliges States "first of all" to settle their disputes by negotiation—provides the foundation stone of the Court's decision. If the Judgment is placed on this high ground, then its rejection of Judge Tanaka's and Judge Morelli's argument that equidistance–special circumstances is a necessary corollary of the natural prolongation principle becomes acceptable.[43] But if Article 33 of the Charter provides the source of the negotiation half of the jointly governing principles, then what is the provenance of the equity half? It is submitted that this is to be found in Article 2 of the Charter, especially in the guarantees of equality and justice.[44]

42. *See* note 18 *supra* and accompanying text.
43. *See* note 18 *supra* and accompanying text.
44. U.N. CHARTER art. 33, paras. 1, 3. It is interesting to note, for example, that the Court remarked: "[T]he use of the equidistance method would frequently cause areas which are the natural prolongation or extension of the territory of one State to be attributed to another, when the configuration of the latter's coast makes the equidistance line swing out laterally across the former's coastal front, cutting it off from areas situated directly before that front." [1969] I.C.J. at 31–32. The Court further stated:

"(a) The slightest irregularity in a coastline is automatically magnified by the equidistance line as regards the consequences for the delimitation of the continental shelf. Thus it has been seen in the case of concave or convex coastlines that if the equidistance method is employed, then the greater the irregularity and the further from the coastline the area to be delimited, the more unreasonable are the results produced. So great an exaggeration of the consequences of a natural geographical feature must be remedied or compensated for as far as possible, being of itself creative of inequity.

"(b) In the case of the North Sea in particular, where there is no outer boundary to the continental shelf, it happens that the claims of several States converge, meet and intercross in the localities where, despite their distance from the coast, the bed of the sea still unquestionably consists of continental shelf. A study of these convergences, as revealed by the maps, shows how inequitable would be the apparent simplification brought about by a delimitation which, ignoring such geographical circumstances, was based solely on the equidistance method." *Id.* at 49 (*see* Map 2).

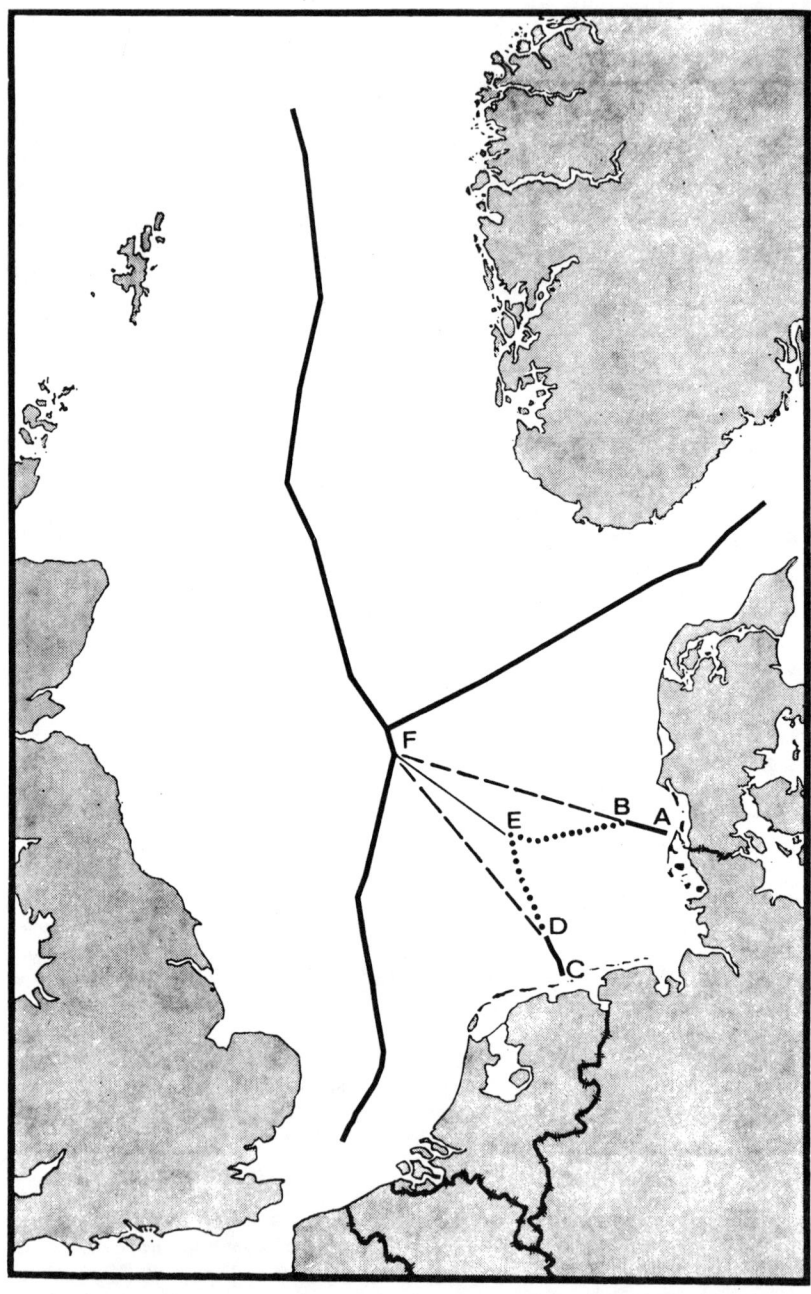

2. Boundary claims of the Netherlands, Federal Republic of Germany, and
   Denmark ([1969] I.C.J. 3, at 16)

The Significance of the Duty to Negotiate—a Possible Analogy

Judge Morelli's view of Article 6 was not limited, as was that of the Court, to its conventional character. For him, the Article restated the customary international law of the shelf as much as did Articles 1–3. It is interesting, therefore, to note that he treated the Article's references to agreement more perfunctorily than the Court did. No doubt he reflected the traditional common-sense approach that looks for agreement without imposing duties of good-faith bargaining.[45] The latter requires that the parties must "not merely go through a formal process of negotiation" but must "so . . . conduct themselves that the negotiations are meaningful."[46] Judge Morelli's downgrading of the duty to negotiate is reminiscent, to some degree, of the difficulties with which the courts in the United States have met in their development of the "duty to bargain in good faith" requirement under Sections 8(a)(5) and 8(b)(3) of the National Labor Relations Act.[47] Again, the Court's insistence that the parties' negotiations should be "meaningful" is reminiscent of the old National Labor Board statement that "[t]rue collective bargaining involves more than the holding of conferences and the exchange of pleasantries . . . . While the law does not compel the parties to reach agreement, it does contemplate that both parties will approach the negotiations with an open mind and will make a reasonable effort to reach a common ground of agreement."[48]

45. Judge Morelli stated: "In fact, in referring to agreement, Article 6 simply means that the States concerned are always free to delimit the continental shelf, by means of an agreement, in the way they think most appropriate, even so as to modify, if appropriate, the existing situation resulting from the application of the equidistance rule. It is to this rule that there must be attributed, even under Article 6, logical and chronological priority.

"When it mentions agreement first, Article 6 adopts the point of view of a court, or of any person or body who proposes to determine the existing legal situation. In order to do this, it is necessary in the first place to ascertain whether an agreement has been concluded by the States concerned. If this is the case, there is nothing to do but hold such agreement to be decisive, because the situation prior to the agreement, and resulting from the equidistance criterion, is no longer in force. It is only in the absence of agreement that the equidistance rule must be applied, by finding for the apportionment effected by that rule, which has not been modified by an agreement." *Id.* at 205–06.

46. *See* text accompanying note 41 *supra.*

47. 29 U.S.C. §§158–59 (1970). *See, e.g.,* Smith, *The Evolution of the "Duty to Bargain" Concept in American Law,* 39 MICH. L. REV. 1065 (1941).

48. Connecticut Coke Co., N.L.B. pt. 2, 88, 89 (1934). This formulation is comparable to Justice Frankfurter's in NLRB v. Truitt Mfg. Co., 351 U.S. 149, 154–55 (1956) where he said: " 'Good faith' means more than merely going through the motions of negotiating; it is inconsistent with a predetermined resolve not to budge from an initial position. But it is not necessarily incompatible with stubbornness or even with what to an outsider may seem unreasonableness."

The Transition from Subjective Good Intentions
to Objective Standards

The United States analogy points to a development in the sophistication of the duty to bargain in good faith. No longer is the subjective element of good faith the touchstone for exculpating a party from censure or sanctions, or for inculpating him. In this vein a leading author in the field of American labor law has written: "Activities which originally were regarded as some evidence of fact carrying legal consequences—in this case bad faith in a truly subjective sense—often comes to be sufficient proof standing alone, thus giving rise to new rules of conduct."[49]

Thus, it is possible to evaluate overt conduct without reference to the party's intent or subjective good or bad faith. But evaluation requires standards whose breach can be judged as bad faith per se. These have been supplied by the National Labor Relations Board[50] and the federal courts in the United States.[51] It is submitted that the rules the International Court of Justice laid down in the *North Sea Continental Shelf Cases* could have a similar function.[52] Their violation would constitute bad faith per se. Again, the Court provided the parties with guidelines for the application of its rules and required the negotiating parties to include the following factors in arriving at the requisite delimitations:

(1) the general configuration of the coasts of the Parties, as well as the presence of any special or unusual features;
(2) so far as known or readily ascertainable, the physical and geological structure, and natural resources, of the continental shelf areas involved;
(3) the element of a reasonable degree or proportionality, which a delimitation carried out in accordance with equitable principles ought to bring about between the extent of the continental shelf areas appertaining to the coastal State and the length of its coast measured in the general direction of the coastline, account being taken for this purpose of the effects, actual or prospective, of any other continental shelf delimitations between adjacent States in the same region.[53]

49. Cox, *The Duty to Bargain in Good Faith*, 71 HARV. L. REV. 1401, 1433 (1958).
50. *See, e.g.,* Truitt Mfg. Co., 110 N.L.R.B. 856, 35 L.R.R.M. 1150 (1954), *enforcement denied,* 224 F. 2d 869 (4th Cir. 1955), *rev'd,* 351 U.S. 149 (1956).
51. *See, e.g.,* NLRB v. Katz, 369 U.S. 736, 743 (1962), where the Supreme Court said that the duty to bargain collectively "may be violated without a general failure of subjective good faith; for there is no occasion to consider the issue of good faith if a party has refused to negotiate *in fact* . . . about any of the mandatory subjects." *See also* NLRB v. General Elec. Co., 418 F. 2d 736 (2d Cir. 1969).
52. *See* text accompanying note 41 *supra.*
53. [1969] I.C.J. at 53–54.

The parties, it is submitted, are obliged to defer to such guidelines in applying the rules the Court formulated in order to discharge their obligations under Article 33 of the United Nations Charter as well as under the Judgment.

The Obligation to Negotiate and the Search for "Yessable Propositions"—A Contrast

If the duty to negotiate, rather than the equidistance principle, provides the necessary means for delimiting coastal States' appurtenant continental shelves, then may there not be a danger that a search on the part of one side for what Professor Roger Fisher[54] has denominated a "yessable proposition" will lead to an unequal bargaining situation? One State may become compelled to concede more than equity would require. Herman Kahn points to an analogous danger:

In those situations in which there is not sufficient mutual interest to strike an acceptable bargain, the net thrust of Mr. Fisher's insights is likely to give excessive and rather effective ammunition to those urging concession and compromise. That is, in many ways Mr. Fisher's recommendations can be used to generate psychological pressures, arguments, and even misleadingly seductive and seemingly neutral observations that are actually recommendations for making concessions and compromises—or even more important, creating the conditions for such concession and compromise. For this reason politicians, the humanists, idealists, utopians and the amateur citizens are going to find this book more sweepingly persuasive than many of the ideologically committed or even some of the relatively hard-headed and tough-minded (in the William James sense) bargainers.[55]

If these comments can be applied to the Court's imposition of a duty to negotiate, then one might sadly observe that the Court's view of equity is merely the Panglossian "whatever is, is best." Such a comment, however, would not take the whole of the Court's thesis into account. By formulating its rules and guidelines, the Court adjusted the balance of negotiating power between the parties and assured them that the strongest or hardest-headed could not assert excessive claims in the name of its concept of the relevant equities nor have leverages for overbearing the others. The Court's formulation of the equitable principles to govern negotiating conduct[56] and the guidelines of fac-

54. R. FISHER, INTERNATIONAL CONFLICT FOR BEGINNERS 15–26 (1969).
55. Kahn, Book Review, N.Y. Times Nov. 9, 1969, §7 (Book Reviews), at 73.
56. [1969] I.C.J. at 47. *See also* text accompanying note 41 *supra.*

tors[57] it laid down for the parties to take into account set formal parameters to the negotiations. The negotiations are neither free-ranging and at large nor so dependent on the strength and will of the stronger or harder-headed as would be the inevitable, if unintended and unforeseen, outcome of Fisher's proposal. The Court's equitable rules and guidelines were formulated so as to govern the North Sea situation; but they are, *mutatis mutandis,* also relevant to questions concerning the settlement of the lines of demarcation between the joining continental shelves of States which abut, either as opposite or adjacent neighbors, on a common continental shelf feature.

The Negotiated North Sea Settlement

On January 28, 1971, Denmark, the Federal Republic of Germany, and the Netherlands concluded their tripartite negotiations for the delimitation of their continental shelf areas under the North Sea. The written agreements consisted of a Protocol[58] signed by all three States and two treaties—between the Federal Republic of Germany and Denmark[59] and between the Federal Republic and the Netherlands.[60] The agreed boundaries extend considerably beyond the area left over after the Dutch and the Danes had each set out their claims[61] in an Agreement of March 31, 1966.[62] This earlier Agreement had been framed partly in terms of the equidistance principle and partly on the basis of the parties being "opposite" States within the terms of Article 6, paragraph 1, of the Continental Shelf Convention[63]—a status denied them by the Court.[64] On the other hand, the area allocated to the Federal Republic was not so extensive as the area claimed by it under its proposed application of the sector principle.[65] The 1971

---

57. *Id.* at 54. *See also* text accompanying note 53 *supra.*
58. 10 Int'l Legal Materials 600 (1971).
59. *Id.* at 603.
60. *Id.* at 607.
61. *See* Annex to Protocol, Agreements Between Denmark, the Federal Republic of Germany, and the Netherlands Delimiting the Continental Shelf in the North Sea, *done* Jan. 28, 1971, 10 Int'l Legal Materials 600, 602 (1971).
62. *See* lines AFC on Map 3 of the Judgment, North Sea Continental Shelf Cases, [1969] I.C.J. at 16 (*see* Map 2). *See also* Map 1 [1969] I.C.J. at 15 (*see* Map 3); Memorial of the Federal Republic of Germany, 1 Pleadings 13, 80–84, 85 (Fig. 21) (*see* Map 4).
63. *See* lines ABEDC on Map 3 of the Judgment, North Sea Continental Shelf Cases, [1969] I.C.J. at 16 (*see* Map 2). *See also* Map 1 [1969] I.C.J. at 15 (*see* Map 3).
64. *Id.* at 28.
65. *See* Memorial of the Federal Republic of Germany, 1 Pleadings 83–86.

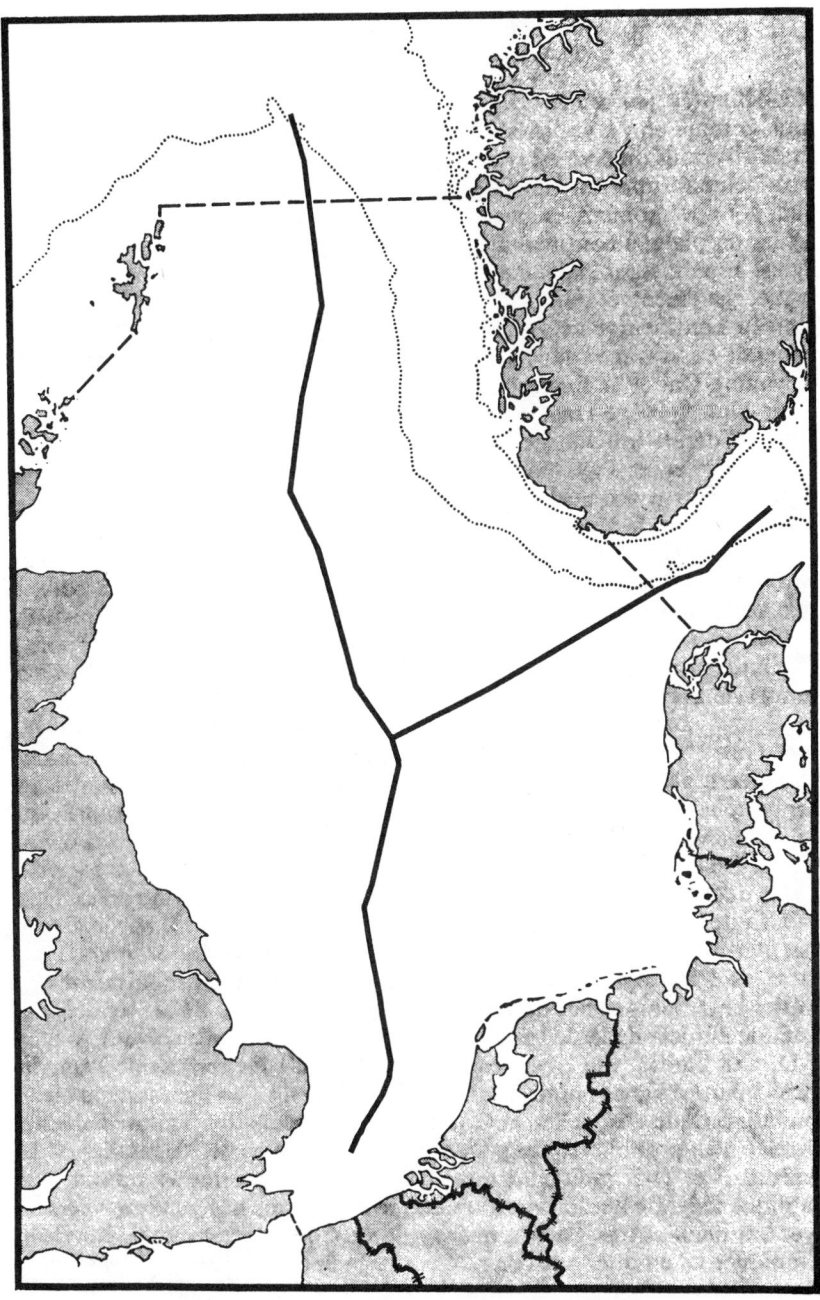

3. Continental shelf delimitations of the United Kingdom, Norway, Denmark, and the Netherlands ([1969] I.C.J. 3, at 15)

Agreement represents more than an attempt to reach a compromise between the Federal Republic's claim and the area left to it by the Danish-Dutch Agreement of March 31, 1966. It reveals an attempt to allocate to each State what should be viewed as being its own continental shelf—this being constituted by the "natural prolongation" of its landward territory.[66] Furthermore, the Court's equitable rules and guidelines can be seen as governing both the demarcation lines and the principles the Agreement prescribes for the exploitation of common deposits existing in the shelves of two or more of the parties.[67]

In the light of the preservation of the traditional freedom to lay cables and pipelines in Article 4 of the Continental Shelf Convention, Article 3 of both of the treaties is of interest. It provides:

Notwithstanding international regulations concerning the laying of pipelines on the continental shelf, pipelines that are laid on the continental shelf in connexion with the exploitation of mineral resources shall in order to prevent the pollution of the sea and to avert any other hazards be subject to the legal provisions relating to the installation and operation of pipelines that are applicable in the territory of the Contracting Party upon whose continental shelf such pipelines are laid.

This clause is conceived of as vindicating, at least in part, the world community interest in environmental protection and enhancement. It requires that the parties to the agreement subject their liberty to lay pipelines, under general law, to the restraints imposed by the domestic laws of each continental shelf State on its own enterprises.

The Saudi Arabia–Iran Agreement

The upper regions of the Persian Gulf, where the opposite Saudi Arabian and Iranian coasts are separated by distances that vary, approximately, from 95 to 135 miles, constitute a shallow marine area. The deepest parts are no more than 250 feet (or some 75 meters), and generally the average depth is far less. It is, like the North Sea, an area that entirely conforms to the Convention's criteria for establishing continental shelf rights as natural prolongations under the sea of the coastal States' land territories.[68] It is, moreover, located in

---

66. [1969] I.C.J. at 47 (the natural prolongation theory as part of the equitable rules applicable in delimiting States' adjacent continental shelves). *See also* note 14 *supra* and accompanying text for the significance of this term to the majority of the Court.

67. Article 2 of each treaty. (While not identical, these two articles are very similar). *See* 10 INT'L LEGAL MATERIALS 600, 604, 608 (1971).

68. *See* note 14 *supra*.

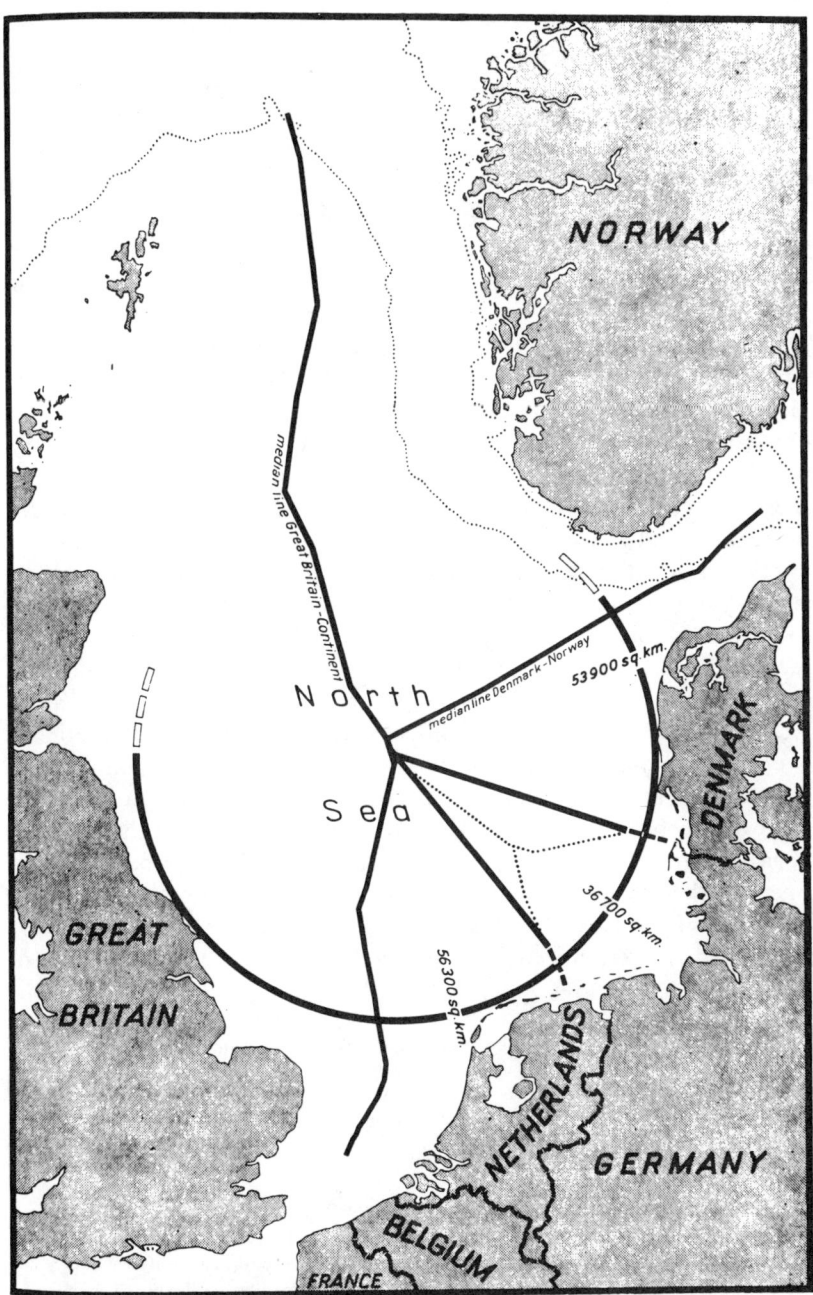

4. Sector principle applied to the North Sea States [Memorial of the Federal
Republic of Germany, 1 North Sea Continental Shelf Cases, I.C.J. Pleadings 85 (1968)]

what has frequently been held to be the world's richest undersea oil region. In addition, offshore oil exploration has indicated the probable existence of large reserves lying well towards the center. The problems of the allocation of the resource and the demarcation of the States' seabed areas were complicated by the presence of a number of islands lying well toward the middle of the Gulf that had been the subjects of territorial disputes and the stationing of garrisons.[69] In light of these facts, it becomes almost inevitable to expect that the issue of the delimitation of the two States' continental shelves was fraught with contentions—even to the use of naval forces to protect concessionaires. But on October 24, 1968, the parties signed an agreement dividing up the seabed area on the basis of equitable principles.[70]

Richard Young has related that the "task of defining the boundary was found to fall into three geographical segments."[71] In the first of these categories, the median line provided the appropriate line of demarcation. In the second, the principal median line was departed from in favor of local median lines and territorial waters limits referable to islands. In the third sector, the demarcation was based on an attempt to make an equitable division of the oil in place on the basis of the technical information available to both sides.[72] Clearly, this demarcation departs substantially from Article 6 of the Continental Shelf Convention. But, as a successful example of good-faith bargaining overcoming considerable emotional as well as economic demands, it reflects the practicality of the decision the Court was to give the following February. The fact that the parties abandoned their earlier positions and agreed on the demarcations of their respective continental shelves in view of equitable principles illustrates the importance of the Court's decision in the *North Sea Continental Shelf Cases* and the relevance of its approach in that case to other diverse regions.[73]

---

69. The facts outlined in this background are largely drawn from Young, *Equitable Solutions for Offshore Boundaries: The 1968 Saudi Arabia–Iran Agreement,* 64 Am. J. Int'l L. 152 (1970).
70. 8 Int'l Legal Materials 493 (1969).
71. Young, *supra* note 69, at 153.
72. *Id.* at 154.
73. Thus, Young has pointed out that "the Saudi-Arabian–Iranian accord appears to be a satisfactory and responsible disposition of a problem in which conflicting interests could have gotten out of hand. In this regard it conforms fully with the principles of international law since declared by the Court to be applicable to such situations, notably the concept of equity and the propriety of an agreed division of any overlapping areas." *Id.* at 156–57. Young saw this agreement as "a valuable precedent for the delimitation of other offshore boundaries between Iran and Iraq,

The Essential Unity of a Deposit of Oil and Gas

Measures for protecting oil deposits from wasteful practices,[74] such as legislation in the United States,[75] Canada,[76] and the United Kingdom[77] (as well as in many other countries), reflect a policy of "unitization,"[78] which is known in the United States as the policy of maintaining the "essential unity of a deposit of oil and gas."[79] In addition to these domestic law examples, the United Nations Secretariat has promoted the policy goal of the "essential unity of a deposit."[80]

Kuwait, the Neutral Zone, and the various smaller states now associated in the Federation of Arab Amirates. Many of these unresolved boundaries present unusually complex problems with respect to islands, baselines, and the division of known or suspected resources; but at least the possibility of amicable agreement has now been demonstrated." *Id.* at 157. In a footnote to this text, Young indicates that "[n]egotiations regarding several of these boundaries are now under way, and in some cases . . . are reportedly well advanced." *Id.* n. 18.

74. Deriving from the application of the "Rule of Capture" to the winning of oil and gas. *See* Westmoreland & Cambridge Natural Gas Co. v. DeWitt, 130 Pa. 235, 18 A. 724 (1889). *See also* Walls v. Midland Carbon Co., 254 U.S. 300 (1920). For a classic statement of the "Rule of Capture" and its correlative offset rule, see Barnard v. Monongahela Natural Gas Co., 216 Pa. 362, 65 A. 801 (1907).

75. 30 U.S.C. §181 (1970). *See* §17b, *added* 1946, 60 Stat. 950, 952, *as amended* 68 Stat. 583, 584–85 (1954) *and* 74 Stat. 781 (1960) [30 U.S.C. §226(j) (1970)]. *See also* Mineral Leasing Act for Acquired Lands, 30 U.S.C. §§351–59 (1970). For a history of the development of §17(b) and its social background, see R. MYERS, THE LAW OF POOLING AND UNITIZATION 294–97 (1957). The Federal Regulations pursuant to the Act are the Unit Plan Regulations, 43 C.F.R. §§192.20, 192.21 (1951). *See also* Suggested Standard Forms, *id.*

76. *See, e.g.,* Oil and Gas Conservation Act of 1955, SASK. STAT. c. 88, §35; Oil and Gas Resources Conservation Act of 1952, REV. ALTA. STAT. c. 46.

77. Great Britain, Petroleum (Production) (Continental Shelf and Territorial Sea) Regulations 1964, Schedule 2 (Model Clauses for Production Licenses), cl. 19, STAT. INSTR. 1964, No. 708.

78. *Unitization* is the term used in American oil and gas law to reflect the idea of the "essential units of a deposit of oil and gas." It is distinct from the idea of "pooling" (which means the "bringing together of small tracts sufficient for the granting of a well permit under applicable spacing rules"), H. WILLIAMS & C. MEYERS, OIL AND GAS LAW §901, at 2 (1964). "Unitization" means "the joint operation of all or some part of the producing reservoir." *Id.* at 3.

79. For the provenance of this, and especially for the U.N. Secretariat's use of it early in the thinking and discussion of the utilization of the resources of a continental shelf region that are to be found on both sides of an international boundary line, see U.N. Secretariat Memorandum on the Regime of the High Seas 109, [1950] 2 Y.B. INT'L L. COMM'N 67, 112, U.N. Doc. A/CN. 4/32 (mimeo. July 14, 1959).

Since the publication of the Memorandum, the term has tended to become as much the accepted term in international law as *unitization* has become for an analogous private law situation in American oil and gas municipal law.

80. *Id.*

Support for this policy is also reflected in authoritative writings[81] and in at least six international agreements.[82] This concept has, however, been given different interpretations. Thus, Mouton has written: "We believe that the principle mentioned in the Memorandum that a dividing boundary-line should not cross an oil-pool and applied in a germane way in the Grisbadarna arbitration, is a guide for countries in framing their delimitation agreements or for the arbitrator who is called in the case of dispute."[83]

In the *Grisbadarna Case,* the Permanent Court of Arbitration did not recognize that Norway might have an equity remaining in the Grisbadarna Banks.[84] Such an equity might well have been capable of

---

81. *See, e.g.,* Memorandum, *supra* note 79, at 109; Lauterpacht, *Sovereignty over Submarine Areas,* 27 Brit. Y.B. Int'l L. 376, 410 n. 4 (1950); Mouton, *The Continental Shelf,* 85 Recueil des Cours 345, 421–23 (Hague Academy of International Law) (1954–I).

   This writer, however, feels compelled to doubt Lauterpacht's statement: "It will be noted that the phenomenon of common pools independent of political boundaries is relied upon in the recitals of the Proclamation of the United States of 1945." Lauterpacht, *supra,* at 410 n. 4. There would appear to be no direct reference to such a phenomenon in the recital to which Lauterpacht refers. Furthermore, the federal policy in these matters, as embodied in §17(b) of the Mineral Leasing Act for Acquired Lands, 30 U.S.C. §226(j) (1970), was not added to the domestic legislation until the year following President Truman's Continental Shelf Proclamation in 1945.

   For an early international arbitration (in 1909) recognizing the principle of the unity of a resource (crustaceans), see Grisbadarna Case (Norway v. Sweden), 13 U.N.R.I.A.A. 147, Hague Courts Reports (Scott) 121 (Perm. Ct. Arb. 1911), 4 Am. J. Int'l L. 226 (1910) (Perm. Ct. Arb. 1909).

82. *See* Agreement between Great Britain and the Netherlands relating to the Exploration of Single Geological Structures extending Across the Dividing Line on the Continental Shelf under the North Sea, [1967] Gr. Brit. T.S. No. 24 (Cmnd. 3254). *See also* note 32 *supra* and the Agreement therein cited; Agreement between Great Britain and Denmark relating to the Delimitation of the Continental Shelf of the two Countries, art. 4, *signed* Nov. 25, 1971, [1973] Gr. Brit. T.S. No. 6 (Cmnd. 5193); Agreement between Great Britain and the Federal Republic of Germany relating to the Delimitation of the Continental Shelf under the North Sea between the Two Countries, arts. 3 and 4, *signed* Nov. 25 1971, [1973] Gr. Brit. T.S. No. 7 (Cmnd. 5192); Agreement between Great Britain and Norway relating to the Delimitation of the Continental Shelf between the Two Countries, art. 4, *signed,* Mar. 10, 1965, [1965] Gr. Brit. T.S. No. 71 (Cmnd. 2757); Agreement between Saudi Arabia and Iran [delimiting the boundary between their respective submarine areas in the Persian (Arabian) Gulf], art. 3, *signed* Oct. 24, 1968, 8 Int'l Legal Materials 493, 494–95 (1969). Surprisingly, Article 2 of the 1968 agreement between Italy and Yugoslavia with respect to the seabed does not qualify that treaty to be brought within the category of treaties cited in this footnote. *See* 1 North Sea Continental Shelf Cases, 1 I.C.J. Pleadings 559, 561–62 (1968).

83. Mouton, *supra* note 81, at 422.

84. *Id.* Additionally, Admiral Mouton said: "The arbitrators held that the boundary-

transformation into a claim for compensation out of a proportion of the Swedish catch. Instead, Sweden was awarded a monopoly of the Grisbadarna lobster fishery, and Norway was given a monopoly of the lobster fishery of the Skjöttegrunde.

This trenchant method[85] of division may be contrasted with the first two Articles of the United Kingdom–Netherlands Single Structure Agreement,[86] which provide:

Article I

If any single geological mineral oil or natural gas structure or field extends across the dividing line and the part of such structure or field which is situated on one side of the dividing line is exploitable, wholly or in part, from the other side of the dividing line, the Contracting Parties will seek to reach agreement as to the manner in which the structure or field shall be most effectively exploited and the manner in which the costs and proceeds relating thereto shall be apportioned, after having invited the licensees concerned, if any, to submit agreed proposals to this effect.

---

line ought to be traced: 'so that it would pass midway between the Grisbadarna banks on the one side and Skjöttegrunde on the other.' The arbitrators did not want to cut through a bank, which is in fact also an application of the same principle of leaving intact the unity of a deposit, this time not of minerals but of marine resources. A demarcation, so the award continues, which would assign the Grisbadarna to Sweden is supported by the circumstance that 'lobster fishing in the shoals of Grisbadarna had been carried on for a much longer time, to a much larger extent, and by a much larger number of fishermen by the subjects of Sweden than by the subjects of Norway.' . . . .

"On the other hand it was averred that 'the Norwegian fishermen have almost always participated in the lobster fishing on the Skjöttegrunde in a comparatively more effective manner than at the Grisbadarna,' which warranted a demarcation assigning the Skjöttegrunde to Norway." *Id.*

85. The Treaty Between the German Reich and the Netherlands for the Determination of the Working Boundary of the Coal Mines Situated on Both Sides of the Frontier Along the River Worm, May 17, 1939, [1939] Staatsblad van het Koninkrijk der Nederlanden No. 30, 199 L.N.T.S. 251, provides another interesting variation on the idea of the unitized working of a mineral deposit. It sought to establish the most economic mining of a common coal deposit on either side of the two countries' frontiers. The solution was that the boundary between the two countries on the surface was not to be applied in the coal galleries. Instead, the treaty provided privileges of continuing a single working from either side of the boundary. There was no recognition of an obligation to compensate the State in whose territory a deposit was to be worked in any given case. Presumably, the thought was that the mutual equities of the States would cancel one another out as they both intended to cross each other's boundaries as the exigencies of mining demanded. This variation, however, although of interest from the comparative law point of view, is clearly not suitable for the rational working of a submarine oil or gas deposit.

86. *See* note 82 *supra.*

Article 2

Where a structure or field referred to in Article 1 of this Agreement is such that failure to reach agreement between the Contracting Parties would prevent maximum ultimate recovery of the deposit or lead to unnecessary competitive drilling, then any question upon which the Contracting Parties are unable to agree concerning the manner in which the structure of field shall be exploited or concerning the manner in which the costs and proceeds relating thereto shall be apportioned, shall, at the request of either Contracting Party, be referred to a single Arbitrator to be jointly appointed by the Contracting Parties. The decision of the Arbitrator shall be binding upon the Contracting Parties.

Like Mouton's proposal, these treaty provisions may make possible the unified working of a single resource or deposit. But, unlike his suggestion, they also call for an apportionment of the costs and proceeds. This apportionment could be measured by reference to the proportion that each State's share of the common deposit bears to the whole. Once this is settled, one State could become the agent State of all the others, and make its distribution of burdens and benefits in terms of the measured or agreed upon apportionment.[87]

The two treaties signed by the Federal Republic of Germany with the Netherlands and with Denmark are also relevant. While they leave open the possibility of a unitized working of a single deposit through negotiation, they emphasize that any agreement the parties negotiate should take "into account the interests of both Contracting Parties on the principle that each Contracting Party is entitled to the mineral resources located on its continental shelf."[88] Furthermore, the second Article in both treaties emphasizes settlement by arbitration rather than by resorting to the wider equities of either negotiation or conciliation.

---

87. For Mouton's proposal, see M. MOUTON, *supra* note 11, at 422. For a discussion of the concept of "agent state," *see* Goldie, *The Oceans' Resources and International Law—Possible Developments in Regional Fisheries Management,* 8 COLUM. J. TRANSNAT'L L. 1, 44–45 (1969).

88. *See* Article 2, paragraph 2, of the Treaty Between the Federal Republic of Germany and Denmark Relation to the Delimitations of the Continental Shelf Under the North Sea, *done,* Jan. 28, 1971, 10 INT'L LEGAL MATERIALS 603, 604 (1971). *See also* Article 2, paragraph 2, of the Treaty Between the Netherlands and the Federal Republic of Germany on the Delimitation of the Continental Shelf Under the North Sea, *done* Jan. 28, 1971, 10 INT'L LEGAL MATERIALS 607, 608 (1971), where the same principle was asserted, in slightly different language, as that quoted in the text. These two agreements, which were signed simultaneously, were accompanied by a Protocol executed at the time by all three parties, 10 INT'L LEGAL MATERIALS 600 (1971).

The Shelf's Outer Limits

The outer limits of the submarine regions over which coastal States may exercise their continental shelf rights are controlled by the criteria provided in Article 1 of the Continental Shelf Convention. These are: (a) the 200-meter bathymetric contour line or isobath; (b) adjacency; (c) the necessary outer limit inherent in the meaning of the term *continental shelf;* and (d) the exploitability test.

The 200-Meter Isobath

Several different definitions of the continental shelf in terms of the configuration of the ocean bottom, rather than man's capabilities to exploit it, have been suggested by various writers. For example, the geographer J. Bourcart defined the world's continental shelves as the submarine landmasses that lie beneath the shallow sea areas between the shores and the *rupture de la pente,* or "break of slope," that is, the first substantial falloff, whatever the depth. Bourcart has also described this zone as the "ocean rim."[89]

Bourcart's concept is predicated on a vision of the continental landmasses and insular areas beyond the continental abyss. Lying offshore but underneath the oceans and between the shores and the depths, a shallow shoulder extends in many places for some distance seaward. This terminates in a steeper slope and plunges into the ocean depths. At the point where the slopes become steeper and pitch downward, Bourcart saw the terminating point of the continental shelf. His definition seeks to reflect geographical reality. Other geographers, noting that despite local variations[90] the average depth of the break in slope tends to approximate that of the 200-meter, or the 100-fathom, bathymetric contour line,[91] have proposed one or the other of these

---

89. In submitting the view that the continental slopes can be more flexibly conceived of as the ocean rim, Bourcart writes: "La rupture du pente séparant le Plateau du Rebord semble constante sur tout le pourtour des mers. Quelle que soit la réduction du Plateau, elle peut suffire à le définir." J. BOURCART, GÉOGRAPHIE DU FOND DES MERS, ÉTUDE DU RELIEF DES OCÉANS 126 (1949).

90. In reality the continental slopes may begin at any depth. In some parts of the world the break in slope occurs at only 50 fathoms, in others not until the depth of 200, or even 500, fathoms is reached.

91. For a treatment of these two isobaths, *i.e.,* the 200-meter and the 100-fathom contour line, as being approximately equivalent, see Int'l L. Comm'n, Commentary to the Articles Concerning the Law of the Sea, [1956] 2 Y.B. INT'L L. COMM'N 265, 295, 11 U.N. GAOR Supp. 9, at 41, U.N. Doc. A/3159 (1956).

two isobaths as the determinant of the shelf's outer limits. They seek to establish one (or even both, as alternatives) of these lines as a criterion having no reference to any physically existing break in slope but offering the advantage of a standardized, if abstract, working definition. The 200-meter isobath has also received the support of the International Committee on the Nomenclature of Ocean Bottom Features of the International Association of Physical Oceanography at Monaco (a member of the International Council of Scientific Unions).[92]

Following the majority of oceanographers, the International Law Commission proposed in 1956,[93] and the 1958 Geneva Conference on

---

There may, however, be a considerable variation of lateral distance on the bed of a gently sloping continental shelf area. Since a meter is approximately 39.37 inches, it may be seen that the 100-fathom isobath is less deep than the 200-meter isobath; the former is 7,200 inches while the latter is 7,874 inches. This difference of 56 feet 2 inches (or just over 2 feet more than 9 fathoms) means that the 100-fathom bathymetric contour line will always be within that of the 200-meter line on the seabed. Depending on the degree of the submarine slope down into deeper zones, and its general configuration, any equivalence laterally on the surface of the continental shelf or slope region between them can only form a rough approximation. Since that possible approximation of the isobaths is so exceedingly rough, the two measures should not be used interchangeably. In this chapter, following the drafts of the International Law Commission, especially that of 1956 and the 1958 Geneva Convention on the Continental Shelf, reference will be made to the 200-meter bathymetric contour line, or isobath, in preference to the 100-fathom line.

92. For the definitions of the continental shelf and related submarine areas accepted by the International Hydrographic Bureau, Monaco, see, 31 INT'L HYDROGRAPHIC REV., May 1954, at 97.

93. Article 1 of the Commission's 1951 Draft Articles adopted the test of exploitability for determining the extent of the "legal" continental shelf, *i.e.,* the formula defined the shelf as being "outside the area of territorial waters, where the depth of the superjacent waters admits of the exploitation of the natural resources of the sea-bed and subsoil." U.N. Doc. A/1858, at 19. In 1951 exploitability provided the sole criterion. It was not collateral with, let alone supplementary to, a criterion depending on a bathymetric contour line. This standard was quickly seen to lack precision and was criticized by many writers, including Boggs, *Delimitations of Seaward Areas Under National Jurisdiction,* 45 AM. J. INT'L L. 240, 245, 265–66 (1951); Feith, *Report,* 44 INT'L L. ASS'N REP. 125, 126–27, 134 (1950); Address by G. Gidel, Fourth Annual Conference of the International Bar Association, July 1952, *translated as* Gidel, *The Continental Shelf,* 3 U.W. AUSTL. L. REV. 87, 89–90 (1954); 45 INT'L L. ASS'N REP. 147 (1952) (Waldock), and a number of governments in their *Comments* on exploitability as the sole test for determining the outer limit of the continental shelf in the 1951 Draft Articles. *See* François, (Fourth) Report on the Regime of the High Seas 17–25, 101–03, U.N. Doc. A/CN. 4/60 (mimeo., Feb. 19, 1953).

As a result of a widespread support for the 200-meter (or 100-fathom) isobath, the Commission adopted that test in Article 1 of its 1953 Draft Articles on the Conthe Law of the Sea accepted, the 200-meter bathymetric contour line in preference to the physical break in slope as one of the criteria for determining, for legal purposes, the outer limits of the continental shelf. In some respects the 200-meter bathymetric contour line—the test in terms of a fixed depth—has considerable advantages over that of the break in slope supported by the geographical realists; it avoids the definitional and legal difficulties attendant upon this latter test's application in concrete situations that admit of contradictory and competing, but equally supportable, claims about the line where the actual break in slope occurs. Such situations may affect wide areas horizontally, when an old and gradually sloping ocean rim is involved.

Adjacency

While the 200-meter bathymetric contour line provides a clear-cut test of what appertains to a coastal State as being within the area of its continental shelf, the Article 1 tests of adjacency and exploitability, which accord to coastal States continental shelf sovereign rights beyond the 200-meter isobath on a contingent basis, are open ended. Indeed, the open-endedness of adjacency can be seen when two comments are compared. First, there is the common-sense comment that "a continental shelf region can remain 'adjacent' to a coast on its landward side and yet be a continuum extending out to mid-ocean areas."[94] (This may be illustrated by reference to the obvious fact that Canada is nonetheless adjacent to the United States despite the vast distance of her Melville Peninsula from the nearest part of the United States). On the other hand, the International Court of Justice has sagaciously pointed out that "it is evident that by no stretch of imagination can a point on the continental shelf situated say a hundred miles, or even much less, from a given coast, be regarded as 'adjacent'

---

tinental Shelf, [1953] 2 Y.B. INT'L L. COMM'N 212, 8 U.N. GAOR Supp. 9, at 12, U.N. Doc. A/2456 (1953). *See id., Comments on the Draft Articles,* para. 61. Parenthetically, it is of interest to note that the exploitability test was eliminated from the 1953 Draft Articles. It was, however, restored by the Commission in 1956. *See* Articles Concerning the Law of the Sea: Part II, Section III, The Continental Shelf, art. 67, [1956] 2 Y.B. INT'L L. COMM'N 256, 264, 11 U.N. GAOR Supp. 9, at 11, U.N. Doc. A/3159 (1956). For a comparative tabulation of the Commission's 1951, 1953, and 1956 Draft Articles, and the Convention on the Continental Shelf, Geneva 1958, see Appendix I to Goldie, *The Contents of Davy Jones's Locker—A Proposed Regime for the Seabed and Subsoil,* 22 RUTGERS L. REV. 1, 58–65 (1967).

94. Goldie, *supra* note 93, at 11. *See also* Goldie, *The Exploitability Test—Interpretation and Potentialities,* 8 NATURAL RESOURCES J. 434, 447–48 (1968).

to it, or to any coast at all, in the normal sense of adjacency, even if the point concerned is nearer to some one coast than to any other."[95]

In many debates on prescribing the outer jurisdictional limits of the continental shelf, this quotation has been used to claim the support of the Court on the side of those who favor the narrow continental shelf over those who favor the wide one.[96] Thus, Wolfgang Friedmann has criticized the American Branch of the International Law Association, E. D. Brown, and R. Y. Jennings by pointing up the divergencies of their various positions from this formula.[97] For example, Friedmann (criticizing Jennings) wrote:

Nowhere does the Court's Judgment[98] support indefinite prolongation of land territory into the sea. It limits national jurisdiction to the continental shelf—a concept clearly distinct from the continental slope and the continental rise—and stresses the concept of "adjacency" which is part of the definition of the continental shelf in Article 1 of the Convention. But Jennings even does away with the latter by stating that "adjacency comprehends the idea of appurtenance as a prolongation of the land domain." Yet the Court has clearly said that "by no stretch of imagination . . . ."[99] For the test of adjacency, Jennings substitutes the test of "exploitability beyond the depths of 200 meters" of

95. North Sea Continental Shelf Cases, [1969] I.C.J. 3, 30.
96. The broad continental shelf may be defined as the extension of the coastal States' exclusive continental shelf jurisdiction out to the continental rise beyond the toe of the slopes where the continental pedestal joins the abyssal plains. This is not really so much a claim to the continental shelf as to the whole continental pedestal. The narrow continental shelf may be defined as the region of coastal States' exclusive continental shelf jurisdiction out to, and terminating at, the 200-meter bathymetric contour line. For a discussion and evaluation of the wide and narrow continental shelf theories, see L. HENKIN, LAW FOR THE SEA'S MINERAL RESOURCES 37–41, 45–46 (ISHA Monograph No. 1, 1968). *See also* Henkin, The Continental Shelf, in Proceedings of the 4th Annual Conference of the Law of the Sea Institute, June 23–26, 1969, at 171, 174–78 (mimeo. March 1970); Henkin, *International Law and "the Interests": the Law of the Seabed,* 63 AM. J. INT'L L. 504 (1969). For the opposite side of this argument, see Finlay, *The Outer Limits of the Continental Shelf,* 64 AM. J. INT'L L. 42 (1970). For examples of assertions of the wide shelf thesis, see Hedberg, Limits of National Jurisdiction over Natural Resources of the Ocean Bottom in Proceedings, *supra,* at 159; Nat'l Petroleum Council, Petroleum Resources Under the Ocean Floor, March 1969, at 55–67 [hereinafter cited as NPC Report]; Nat'l Petroleum Council, Petroleum Resources Under the Ocean Floor: A Supplemental Report, March 1971, *passim;* Comm. on Deep Sea Mineral Resources of the Am. Branch of the Int'l L. Ass'n, Interim Report, July 19, 1968, *passim;* Comm. on Deep Sea Mineral Resources of the Am. Branch of the Int'l L. Ass'n, Second Interim Report, July 1970, at 1, 4–5. A review of these two Interim Reports shows them to be additional manifestations of the National Petroleum Council's anxieties.
97. W. FRIEDMANN, THE FUTURE OF THE OCEANS 39–42 (1971).
98. North Sea Continental Shelf Cases, [1969] I.C.J. 3.
99. *See* text accompanying note 95 *supra* for the full quotation from the Judgment.

Article 1 of the Convention (ignoring that this Article, too, speaks of "areas adjacent to the coast").[100]

Friedmann's animadversions on the definition of adjacency by the Court have also provided Henkin with one of the supporting arguments[101] for his criticism of the National Petroleum Council's advocacy of a wide continental shelf in a series of reports and other manifestations.[102] After reviewing the context in which the Court put forward its formula regarding adjacency, one may well doubt whether the apparently crucial quotation from the Court's Judgment was constructed to support the arguments for which Friedmann and Henkin cite it. When read in context, it really does not appear to be the basic and generally applicable statement those writers apparently characterize it to be.

The problem of the Court's meaning is complicated. It had rejected the applicability of Article 6 of the Continental Shelf Convention[103] in favor of imposing a duty of good-faith bargaining in order to delimit continental shelves among adjacent States and among opposite States that abut on a common continental shelf feature.[104] Does this mean, as both Henkin's and Friedmann's reliance on the Court's use of adjacency necessarily implies, that the Court was defining the term as a modality for applying Article 1 of the Continental Shelf Convention (which, being an embodiment of customary international law in the Court's view, would govern all the parties to the case if applicable)? An affirmative answer might appear to be called for, at least on a superficial appraisal. On the other hand, the Court was dealing with a situation in which the adjacent States (the Netherlands and Denmark

---

100. W. FRIEDMANN, *supra* note 97, at 41–42.
101. Henkin, *"Interests,"* *supra* note 96, at 508.
102. *Id. See* Ely, *American Policy Options in the Development of Undersea Mineral Resources,* 2 INT'L LAWYER 215 (1968), 1 NATURAL RESOURCES LAWYER 91 (1968); Ely, *Current International Issues Relating to the Law of the Sea,* 4 NATURAL RESOURCES LAWYER 569 (1971); Ely, *Draft United Nations Convention on the International Seabed Area: The American Bar Association Position,* 4 NATURAL RESOURCES LAWYER 60 (1971); Finlay, *Draft United Nations Convention on the International Seabed Area: American Petroleum Institute Position,* 4 NATURAL RESOURCES LAWYER 73 (1971); Finlay, *The Outer Limit of the Continental Shelf, A Rejoinder to Professor Henkin,* 64 AM. J. INT'L L. 42 (1970). *See,* however, with respect to this last, Henkin, *"Interests,"* *supra* note 96, at 504.
103. For an analysis of this case's significance, see Goldie, *supra* note 7, at 325, and, more particularly, for a discussion of the Court's rationale for finding Article 6 of the Continental Shelf Convention not applicable to the case before it, see *id.* at 330, 336–38.
104. For a discussion of this introduction of an important new doctrine into public international law, see Goldie, *supra* note 7, at 359–67. *See also supra* at 13–28.

are neither adjacent nor opposite[105] each other, but both are adjacent
to the Federal Republic of Germany) have adjacent continental
shelves, all of which are both the natural prolongations of the
landward territory of their appropriate States and are adjacent to
each other.[106] In addition, as the Court was careful to point out,[107] the
dispute related to a submarine area whose whole extent was above the
depth of 200 meters.[108] Accordingly, the term *adjacency,* as a modality of the outer limits of shelf regions appertaining to States under
the exploitability test, was not in issue. Again, the Court expressed
concern to distinguish equidistance from adjacency. Adjacency should
not be applied to allow a situation where a "lateral equidistance line
often leaves to one of the States concerned areas which are a natural
prolongation of the territory of the other."[109] Here we note the
Court's reference to adjacency to prevent the overlapping of adjacent
States' continental shelves. It saw the purpose of the duty to bargain
continental shelf boundaries in good faith[110] as to ensuring that "the
natural prolongation of a State's land territory must not encroach
upon what is the natural prolongation of the territory of another
State"[111] and to establish the modalities for determining that natural
prolongation.

The Court was confronted by adjacency operating differently in

105. [1969] I.C.J. at 28.
106. For the Court's reliance on the concept of the continental shelf as the natural pro-
    longation of the coastal States' landward territory, see *id.* at 29–31, 36–37, 39–40,
    43–44, 47, 51. For Friedmann's objection to the Court's formulation of the term
    natural prolongation, see Friedmann, *The North Sea Continental Shelf Cases—A
    Critique,* 64 AM. J. INT'L L. 229, 236–37 (1970). On the complexity of the concept
    of adjacency, when related to the continental shelf defined as "the natural prolon-
    gation of the coastal State's land domain" in the context of at least three abutting
    States, see Goldie, *supra* note 7, at 332–36.
107. *See* [1969] I.C.J. at 13, where the Court stated: "The waters of the North Sea are
    shallow, and the whole seabed consists of a continental shelf of less than 200
    metres, except for the formation known as the Norwegian Trough, a belt of water
    200–650 metres deep, fringing the southern and south-western coasts of Norway
    to a width averaging 80–100 kilometres."
108. The only portion of the North Sea that extends to depths greater than 200 meters,
    namely the feature known as the Norwegian Trough, was outside the scope of this
    dispute as being, in the Court's words, "already . . . the subject of delimitation by a
    series of agreements concluded between the United Kingdom . . . and certain
    States on the Eastern side, namely Norway, Denmark and the Netherlands." *Id.*
    at 13–14.
109. *Id.* at 37.
110. Note 7 *supra. See also* paragraph (a) of the quotation from the Judgment of the
    Court in The North Sea Continental Shelf Cases in the text accompanying
    note 41 *supra.*
111. *Id.* at 47.

different dimensions—those of coastal States adjacent to each other, of adjacent shelves appurtenant to coastal States, and of shelves that appertain as adjacent (being natural prolongations) to their own coastal States while remaining physically closer to other coastal States.[112] This legal and geographical complex was set within the closed limits of the North Sea continental shelf areas. The different contexts of the term underscore the relativity of the term *adjacency* as the Court used it in the multidimensional North Sea situation. They also bring one to question whether the Court's denial that continental shelves "situated a hundred miles, or even much less" could be adjacent was relevant for elucidating the meaning of the term outside the intricate situation of the *North Sea Continental Shelf Cases*. The complexity of the circumstances in the case under discussion and the enormous variability of the contexts in which the term can be used cast more doubt on the aptness of citing the Judgment's reference to stated distances,[113] which it formulated for the purpose of a specific fact-pattern, as supporting a generally applicable rule of construction for applying Article 1 of the Continental Shelf Convention in all cases, especially since the interpretation of Article 1 was not an issue in the case before the Court.

In brief, while the Court's reference to distances in defining its notion of adjacency may have been relevant to the special fact-pattern of the *North Sea Continental Shelf Cases,* it should not, for that reason, be viewed as a necessary criterion for interpreting Article 1 wherever it is applied. The distances that the Court mentioned might well be completely irrelevant in cases where the outer limits of the permissible shelf region are being delimited without the restraining presence of adjoining, let alone opposite, States or where the complex interplay of claims is not circumscribed within the confines of an enclosed sea area such as the North Sea.

The submission here, therefore, is that the term *adjacency* as used in Article 1 is a relative one and does not carry with it any necessary maximum distance from the shore. Its relativity is dependent on such limitations of political and physical geography, history, and topography as those in the *North Sea Continental Shelf Cases,* or on the lack of such restricting factors. Adjacency in the North Sea context will tell us little that is directly relevant to the Southeast Pacific, except to distrust its use in the absolute terms in which, for example, Friedmann and Henkin would appear to be uncharacteristically citing it. Indeed, adjacency should not be invoked independent of its context.

112. *Id.* at 31.
113. *See* text accompanying note 95 *supra.*

The legal institution we are discussing is the legally adjacent continental shelf.

Some further light on the meaning of *adjacent* and on its function as indicating what the Convention intended to present as the legal continental shelf is shed by the history of its inclusion. In the International Law Commission's 1951[114] and 1953[115] Drafts of Article 1, the word *adjacent* was not used in defining the continental shelf. The word used to relate the shelf to the coast was *contiguous*. The change was made in Article 67 of the International Law Commission's 1956 Articles Concerning the Law of the Sea.[116] This was carried over into the Continental Shelf Convention's Article 1. The concern that led to the selection of *adjacent* over *contiguous* would appear to have stemmed from the felt ambiguities of the latter term.[117] There was a strong concern to assure the continental shelf rights of States over continental shelves having considerable proximity to their coasts but separated therefrom by narrow channels deeper than 200 meters.[118] The Commission wished to ensure that the presence of an "adjacent continental shelf," whether a geographical feature, such as that found in the case of the Norwegian Trough, or the more common type of off-lying reefs and shallows (of less than 200 meters) forming part of one and the same "three dimensional bench," should give rise to the same legal consequences as a "contiguous continental shelf." The same considerations of policy, convenience, proximity, and the orderly de-

114. Draft Articles on the Continental Shelf and Related Subjects in Int'l L. Comm'n, [1951] 2 Y.B. INT'L L. COMM'N 123, 141, 6 U.N. GAOR Supp. 9, at 17, U.N. Doc. A/1858 (1951).

115. Int'l L. Comm'n, Report, [1953] 2 Y.B. INT'L L. COMM'N 200, 212, 8 U.N. GAOR Supp. 9, at 12, U.N. Doc. A/2456 (1953).

116. Int'l L. Comm'n, Report, [1956] 2 Y.B. INT'L L. COMM'N 253, 264, 296, 11 U.N. GAOR Supp. 9, at 12, 44, U.N. Doc. A/3159 (1956).

117. On the acceptance of these terms as having identical meanings with contiguity constituting the technical term regarding territorial issues, see Goldie, *Australia's Continental Shelf: Legislation and Proclamations,* 3 INT'L & COMP. L.Q. 535, 561-63 (1954), and the authorities there cited.

118. [1956] 1 Y.B. INT'L L. COMM'N 137, U.N. Doc. A/CN. 4/99/Add. 1 (1956) (Dr. Garcia Amador, Chairman, J.P.A. François, Special Rapporteur, and Sir Gerald Fitzmaurice); Commentary (8) on Article 67 (the Commission's Draft of what later became Article 1 of the Continental Shelf Convention); Int'l L. Comm'n, Report, [1956] 2 Y.B. INT'L L. COMM'N 297, 11 U.N. GAOR Supp. 9, at 42, U.N. Doc. A/CN. 4/97 (1956); 6 U.N. CONF. ON THE LAW OF THE SEA, GENEVA 1958, OFFICIAL RECORDS (FOURTH COMM.) 9 (Alvarez Aybar), U.N. Doc. A/CONF. 13/42 (1958); *id.* at 41 [Gutteridge, discussing the joint proposal of the Netherlands and Great Britain, U.N. Doc. A/CONF. 13/C.4/L.32 (1958)]. Note the use of the term *adjacent* in paragraph 1 of the United Kingdom's revised proposal, U.N. Doc. A/CONF. 13/C.4/L.24/Rev. 1 (1958).

velopment of natural resources were seen as applying in areas where depths of more than 200 meters separate the landward seabed from the further seaward zones of the same shelf with as much cogency as they do when the shelf is a continuous and gently sloping feature. Disputes about the recently discovered submarine oil deposits offshore from the Senkaku Islands (thought to be among the world's largest) may test this issue. This small group of islands stands on the Chinese continental shelf in the sense that current charts and surveys show no channels or trenches of more than 200 meters in depth separating them from the shelf area, which is uniformly less than 200 meters in depth. On the other hand, they are far closer to the Ryukyu Islands but are separated from this substantial island chain by a trench of considerably greater depth than 200 meters.[119] Which should govern, absent any agreement between the parties: the criterion of depth or that of adjacency, assuming the Senkakus to be judged more adjacent to the Ryukyu Islands than to Mainland China?

Indeed, much of the difficulty that has arisen with the use of the term *adjacent* in Article 1 of the Convention would be dispelled if commentators took it in context, rather than in the abstract. The Convention, and the customary international law it reflects, deals with continental shelf rights. These rights appertain to coastal States conditionally, and when the geographical feature we know as the continental shelf exists. Any divergencies between definitions of the geographical and the legal shelves should not blind us to the fact that both the International Law Commission and the 1958 United Nations Conference on the Law of the Sea rejected such alternative terms as *submarine areas* precisely because they did not carry any essential connotations of the shelf.[120] This view is confirmed by the *North Sea Continental Shelf Cases.* In stressing the shelf's characteristic as the natural prolongation of the coastal State's land territory, the Court

119. *See, e.g.,* N.Y. Times, Aug. 28, 1969, at 1, col. 7; *id.* at 4, cols. 3–5; *id.* Sept. 1, 1969, at 2, cols. 7–8; *id.* Sept. 24, 1969, at 3, col. 2. *See also* Map 6. For a discussion of this problem see Goldie, *The International Court of Justice's "Natural Prolongation" and the Continental Shelf Problem of Islands,* 4 NETHERLANDS Y.B. INT'L. L. 237 (1973). For a specific discussion of the East China Sea problems of islands, *see id.* at 252–59.

120. For reference to the rejection of a proposal to substitute *submarine areas* for *continental shelf* in Article 1, see Int'l L. Comm'n, Report, [1951] 2 Y.B. INT'L L. COMM'N 123, at 141, 6 U.N. GAOR. Supp. 9, Annex, art. 1, para. 3, Commentary, U.N. Doc. A/1858 (1951); [1953] 1 Y.B. INT'L L. COMM'N 77 (Lauterpacht); [1956] 1 Y.B. INT'L L. COMM'N 132 (Francois); Commentary (9) on Article 67, [1956] 2 Y.B. INT'L L. COMM'N 297, 11 U.N. GAOR Supp. 9, at 42, U.N. Doc. A/3159 (1956); U.N. Doc. A/CONF. 13/42, *supra* note 118, at 35–36 (Gutteridge), 37 (Ruiz-Moreno), 43 (Obiols-Gomez).

emphasized the dependence of the concept of the legal shelf and its legal unity with the coastal State upon the physical unity of the geographical shelf (not, one should note, the geographical pedestal including the slope and the rise) with the dry land area. Surely the supporters of the narrow continental shelf can rest their argument on the meaning of the term *continental shelf* and so resist its confusion with the whole continental pedestal, rather than follow the will-o'-the-wisp of the word *adjacency* in Article 1 of the Continental Shelf Convention.

The Meaning Inherent in The Term Continental Shelf

At its 1951 and 1953 sessions,[121] the International Law Commission was invited to use the term *submarine areas* rather than *continental shelf* since the submarine areas subject to coastal States' sovereign rights might not always be continental shelves in the geographical sense. The Commission declined this invitation on the ground that, despite possible variations from the geographical norm, the term *continental shelf* carried with it connotations that helped to indicate the Commission's intentions and limit the claims of coastal States.[122] The Commission's choice, and the adoption of the descriptive term *continental shelf* by the 1958 United Nations Conference on the Law of the Sea at Geneva, was made with a view to limiting the types of submarine areas that qualified for coastal States' sovereign rights to those showing an effective relationship with the oceanographic and geological notion of the continental shelf. Reflecting this point of view, Gohar, representative of the United Arab Republic to the Fourth Committee at the 1958 United Nations Conference on the Law of the Sea, pointed out that "the continental shelf was not an arbitrary limitation, as was the territorial sea, but a natural feature . . . ."[123] Hence it follows that claims extending the outer limit of the continental shelf based on the exploitability test should still be restricted to what is in fact the continental shelf. This feature is part of the continental pedestal and should not, without abuse to the language of Article 1, be read as engulfing the continental slopes, terraces, toe, and rise. These terms, too, have independent connotations that cannot be conjured away by emphasizing the potentially unlimited scope of the exploitability test.

121. *See* note 120 *supra* for citations to arguments supporting this thesis.
122. *See, e.g.,* [1953] 1 Y.B. INT'L L. COMM'N 76 (Laing), 80 (Amado). *See also* matters cited in note 120 *supra*.
123. U.N. Doc. A/CONF. 13/42, *supra* note 118, at 27.

The Effects of Small Islands and Islets

Small islands and islets may exist offshore from a State claiming a part of a common continental shelf. They may lie out in the center of the shelf area or beyond the median line or may even adjoin the coast of an opposite State on the common shelf. For example, sovereignty over the Channel Islands, including the Minquiers and Ecrehos [Ecrehou] Reefs,[124] vest in the United Kingdom. Those latter groups of islets and reefs come within 6.6 miles of the mainland coast of French Normandy.[125] Should the United Kingdom's continental shelf extend to a median line only 3.3 miles from France?[126] Little has been written bearing directly on this problem, and, apart from one recent article,[127] guidance is scarce on what international law provides for giving effect in continental shelf delimitations to islands' and islets' size, population, level of development, economic and social significance, distance from shore, and location in terms of deposits of submarine resources.

The whole issue of what maritime and continental shelf areas may be considered as appurtenant to islands and islets has been permitted to become a vexed one, since no minimum size and population requirement has been authoritatively set, and the question of geographical continuity with the mainland, or the main island or island group of an island State (for example, Japan or the United Kingdom), has not

124. Sovereignty over these groups of islets and reefs, insofar as they are capable of appropriation, was adjudged by the International Court to belong to the United Kingdom. *See* Minquiers and Ecrehos Case, [1953] I.C.J. 47, 72. The Court described their physical location as follows: "These groups lie between the British Channel Island of Jersey and the coast of France and consist each of two or three habitable islets, many smaller islets and a great number of rocks. The Ecrehos group lies north-east of Jersey 3.9 sea-miles from that island, measured from the rock nearest thereto and permanently above water, and 6.6 sea-miles from the coast of France, measured in the same way. The Minquiers group lies south of Jersey, 9.8 sea-miles therefrom and 16.2 sea-miles from the French mainland, measured in the same way. This group lies 8 sea-miles from the Chausey Islands which belong to France."

125. *See* note 124 *supra* for a description of their adjacency to the coast of France (*i.e.,* French Normandy).

126. Note the distance of the nearest rock that remains permanently above water of the Ecrehos group as described in the quotation from the Minquiers and Ecrehos Case, note 124 *supra*. It would appear that the map (Fig. 13) showing the median line between the Channel Islands and the continent in the Memorial of the Federal Republic of Germany in the North Sea Continental Shelf Cases fails to take into account the possible effects of the Minquiers and Ecrehos features on an Anglo-French median line in the English Channel. *See* 1 Pleadings 48.

127. *See* Ely, *supra* note 2, at 219.

been resolved. This confusion is compounded when Article 1 of the Continental Shelf Convention is read with Article 10, paragraph 1, of the Convention on the Territorial Sea and the Contiguous Zone.[128] Article 1 of the Continental Shelf Convention provides:

> For the purpose of these articles, the term "continental shelf" is used as referring (*a*) to the seabed and subsoil of the submarine areas adjacent to the coast but outside the area of the territorial sea, to a depth of 200 metres or, beyond that limit, to where the depth of the superjacent waters admits of the exploitation of the natural resources to the said areas; (*b*) to the seabed and subsoil of similar submarine areas adjacent to the coasts of islands.

Article 10, paragraph 1, of the Convention on the Territorial Sea[129] provides: "An island is a naturally-formed area of land, surrounded by water, which is above water at high-tide."

In the *North Sea Continental Shelf Cases*, Article 1 of the Continental Shelf Convention was held to enshrine customary international law.[130] On the other hand, a literal application of Article 10, paragraph 1, of the Convention on the Territorial Sea and Contiguous Zone would be at variance with customary international law, since that would confer for the first time territorial seas on small islets. This would be a considerable extension of what has been traditionally recognized. Similarly, if the definition of *islands* in Article 10, paragraph 1, of the Convention on the Territorial Sea and Contiguous Zone were literally applied to determine what islands are included in Article 1 of the Continental Shelf Convention, many paradoxes would result. For example, islands belonging to more distant States, lying close to the coasts of States opposite their parent nation, and abutting on a common continental shelf feature could provide the more distant State with arguments justifying its engrossment of the greater part of the common seabed area. Such an appraisal, in turn, would lead to contradictory applications of the Court's ambiguous test of natural prolongation as each State sought to invoke its aid to maximize the shelf area it could incorporate within the scope of its sovereign rights. Nor, in such a situation, could the phrase "natural prolongation" provide an effective touchstone for resolving those contradictions, except by reference to extraneous criteria. Yet, so it would appear, there is no conclusive authority for denying that islets literally complying with Article 10, paragraph 1, of the Convention on the Territorial Sea and Contiguous Zone (namely, that they remain uncovered by the sea

128. *Done* Apr. 29, 1958, [1964]2 U.S.T. 1606, T.I.A.S. No. 5639, 516 U.N.T.S. 205
     (effective Sept. 10, 1964) [hereinafter cited as Convention on the Territorial Sea].
129. Note 128 *supra*.
130. [1969] I.C.J. at 39.

at high tide) may have their own continental shelves, despite the unreasonable results that would flow from this position. There is,
however, both principle and persuasive precedent for denying that
such islets as those in the Minquiers and the Ecrehos groups are
entitled to provide a basis for claiming either an insular shelf or
asserting a claim to a share of a continental shelf at the expense of
abutting States.

A premise for the argument that follows is that if there are categories of islands and islets not entitled to an appurtenant territorial
sea, the States to which they belong will not be entitled to assert continental shelf claims to adjacent submarine areas. On the other hand,
even if they have been recognized as entitled to claim a belt of territorial sea, their claims to adjacent continental shelves should be more
limited than those that Article 1 would appear to give on a literal interpretation.

It is true that *The Anna* appeared to provide authority for asserting
that even the smallest and most evanescent islands are entitled to their
own territorial seas.[131] Sir William Scott, later Lord Stowell, speaking
for the court, held that a prize could not be taken by a British
privateer within three miles of the mud islets at the mouth of the
Mississippi River, which remained dry at high tide. He judged that
these could provide base lines for measuring the territorial waters of
the United States, since he viewed them as a "kind of portico to the
mainland . . . . and as the natural appendage of the coast on which
they border."[132] This decision apparently has been strongly confirmed
by that of the Judicial Committee of the British Privy Council in the
Secretary of State for India v. Chellikani Rama Rao,[133] as well as the
decisions of many other States in a variety of contexts. All such decisions, however, are supportable only when the islands in question have
the same close relationship with the dry land territory of the State as
in the two cases just cited: where they were formed by alluvion, where
they enjoy close proximity to, or identity with, the mainland of the
coastal State, (for example, the islands of the Norwegian and Scottish
skerries), or where they are located within the geographical unity of
the coastal State or the main islands of an island or archipelago State.
If new islands come into being or are discovered on the high seas, they
belong to no State, but, along with whatever territorial waters are appropriate, they must be acquired by occupation; that act may be
performed by any State.[134] Its appurtenant territorial waters, should

131. 5 C. Rob. 373, 165 Eng. Rep. 809 (1805).
132. 5 C. Rob. 373, 385c, 165 Eng. Rep. 809, 815 (1805).
133. 32 T.L.R. 652 (1916).
134. *See* 1 L. OPPENHEIM, INTERNATIONAL LAW 565 (8th ed. H. Lauterpacht 1955).

the new islet be considered to possess them, are also independent of those appurtenant to the mainland.

In contrast, islets and rocks that remain dry at high tide are almost uniformly considered as part of the high seas, in the absence of such special circumstances as those illustrated in the cases just discussed. The establishment of lighthouses, radio and radar stations, and even fortifications upon them cannot alter their status. The British Foreign Office Legal Adviser's opinion that the Eddystone Rock and Lighthouse does not possess a territorial sea and that French fishing boats are entitled to fish in close proximity and well within three miles of it is one illustration of this point.[135] The recent action taken by the United Kingdom over the islet of Rockall in the North Atlantic provides another illustration. This thesis, finally, is confirmed by the International Court of Justice's observation in the *North Sea Continental Shelf Cases* that "only the presence of some special feature, minor in itself—such as an islet of small protuberance—but so placed as to produce a disproportionately distorting effect on an otherwise acceptable boundary line, would, so it was claimed, possess this character."[136] And later in the Judgment the Court asserted: "The continental shelf area off, and dividing, opposite States, can be claimed by each of them to a natural prolongation of its territory. These prolongations meet and overlap, and can therefore only be delimited by means of a median line; and, ignoring the presence of islets, rocks and minor coastal projections, the disproportionally distorting effect of which can be eliminated by other means, such a line must effect an equal division of the particular area involved."[137]

If, however, a feature should be regarded as too large to be ignored in the process of territorial sea or continental shelf delimitations, there is still no warrant for going to the opposite extreme of awarding it the appurtenances of an island community if it does not qualify as such.

Islands thus create sharp problems qualifying for a special circumstances approach. As Professor Shigeru Oda commented:

Thus, taking into account the existence of all islands in drawing the median line is not conceivable. The existence of islands is no more than one of the factors to justify special circumstances. The State which wishes to claim the special circumstances has to prove them. However, so far as there is no precedent at all in fact, success or failure of proving special circumstances, cannot but depend upon whether or not the mutual consent of the States

---

135. *See, e.g.,* T. FULTON, SOVEREIGNTY OF THE SEA 641–44 (1911). *See also* 1 A. MCNAIR, INTERNATIONAL LAW OPINIONS 371–73 (1956).
136. [1969] I.C.J. at 20.
137. *Id.* at 36.

concerned is obtained unless decided by other means such as international arbitration or the International Court of Justice.[138]

In the *North Sea Continental Shelf Cases,* the International Court of Justice provided guidelines for fixing the parameters of the "special circumstances" when islands impinged on the demarcation of continental shelf delimitations, formulating the "factors" previously discussed.[139]

Northcutt Ely recently surveyed some thirteen instances of contemporary State practice.[140] He found that when small islands would create a distorting effect on a boundary, they have been ignored, but when such islands have a substantial population or economic importance they have been conceded a continental shelf with a twelve-mile radius, equivalent in area to a twelve-sea-mile contiguous zone.[141] Ely concluded his presentation with the following proposal based on contemporary international practices:

> We suggest, therefore, that in these cases in which islets are denied recognition as basepoints for the calculation of a median or equidistance line in the demarcation of seabed boundaries because of their small size and distance from their owner's major territories, their "special circumstances" can be met by recognizing that another State cannot extend its continental shelf rights (that is, rights to exploration and exploitation of submarine resources) into the seabed and subsoil underlying the waters of the islet's contiguous zone, and that only the State owning that islet can explore and exploit the seabed resources which are within its contiguous zone. The result of this reasoning is that an islet which is denied effect as a basepoint for the calculation of a median line as against an opposite coast, or the calculation of an equidistance line between adjacent States on the same coast, for demarcation of continental shelf boundaries, is, nevertheless, to be accorded continental shelf rights in an area of the seabed and subsoil co-extensive with its 12-mile contiguous zone.[142]

Natural Prolongation as a Definitive Criterion

The International Court of Justice's coinage of the phrase "natural prolongation" has been criticized for possible accidental historical

138. Oda, *Boundary of the Continental Shelf,* 12 JAPANESE ANN. INT'L L. 264, 280–82 (1968). *Accord,* M. McDOUGAL & W. BURKE, *The Public Order of the Oceans* 436–37 (1962); Lauterpacht, *supra* note 81, at 410; Gutteridge, *The 1958 Geneva Convention on the Continental Shelf* 35 BRIT. Y.B. INT'L L. 102, 120 (1959); Padwa, *Submarine Boundaries,* 9 INT'L & COMP. L.Q. 628 (1960).
139. *See* text accompanying note 53 *supra.*
140. Ely, *supra* note 2, at 227–30.
141. *Id.* at 227–30.
142. *Id.* at 236.

connotations and for its lack of philosophical felicity,[143] but, it is now part of the positive international law of the continental shelf as the customary law principle reflected in Article 2 of the Convention.[144] Its use complements the Court's policy of emphasizing the legal dependence of the continental shelf on the mainland territory of the coastal State. Thus, the principle asserting that the coastal State's continental shelf is "the natural prolongation of its land territory"[145] subordinates the consideration of small, offshore or dependent islands to that of the mainland, or of the major islands of an island or archipelago State, in any continental shelf demarcation. The natural prolongation of the United Kingdom's continental shelf under the English Channel would not extend within 3.3 miles of the French coast. Rather, the effective natural prolongations of those two countries' shelves should be settled by negotiation, giving due weight to the claims of the communities on the Channel Islands and to the United Kingdom as a whole, which includes those islands, but also recognizing the French equities, including those based on some variant of the maxim "equity is equality."

The Exploitability Test

In addition to the foregoing tests, Article 1 of the Convention provides a further test for delimiting the outer limits of coastal States' continental shelves beyond the 200-meter isobath, namely, "where the depth of the superjacent waters admits of the exploitation of the natural resources of the said areas."[146] This fourth criterion is frequently known as the "exploitability test." The question that arises is whether two of the criteria, or tests, namely, that of the 200-meter bathymetric contour line and that of exploitability, can be taken in conjunction. The history of the exploitability test in the deliberations of both the International Law Commission and the 1958 United Nations Conference on the Law of the Sea at Geneva shows that it was meant to have a supplementary and subordinate function to that of the 200-meter isobath test. Originally it was intended to permit a coastal State to exercise sovereign rights over continental shelf activities carried out in areas beyond the 200-meter isobath and in the continuation of activities begun, or connected with those carried out,

143. *See, e.g.,* Friedmann, *supra* note 106, at 236–37.
144. *See, e.g.,* [1969] I.C.J. at 22, 31–32.
145. *See* note 14 *supra* and the references there cited.
146. The Article 1 definition further identifies these areas as "submarine areas adjacent to the coast but outside the area of the territorial sea."

in the zone between its territorial sea and the 200-meter bathymetric contour line.[147] The test was thus intended to provide a practical solution to day-to-day problems that would arise if the 200-meter bathymetric contour line were accepted as a complete and final cutoff line.[148] Despite the original conception, however, the exploitability test no longer has the merely supplementary function and subordinate status envisaged for it in 1958.

Social Utility and Indeterminacy—The Exploitability Test

Admittedly, apart from the 200-meter isobath test, the criteria for determining the outer limits of the continental shelf of the coastal State for the purposes of the Convention are open ended. On the other hand, the definition of the continental shelf ascribable to a coastal State by reference to the 200-meter bathymetric contour line may seem to have at least a tenuous connection with Bourcart's geographical shelf. For, although in any given case the coincidence of the break in slope with the 200-meter bathymetric contour line will be accidental (since the physical break in slope and the 200-meter isobath can be widely divergent), there would, nevertheless, appear to be a general and perhaps abstract and theoretical congruence. But no designation of a submarine region as a continental shelf, defined in terms of the exploitability test, need coincide with the physical geography of the oceans' contours. That test is quite clearly independent of the

147. On the checkered history of this test in the International Law Commission's Drafts, see note 93 *supra.* On the exploitability test's evaluation by the Commission in 1956 as a subordinate and supplementary concept, see the Commentary on Article 67 (the Commission's Draft of what later became Article 1 of the Continental Shelf Convention), [1956] 2 Y.B. INT'L L. COMM'N 296–97, 11 U.N. GAOR Supp. 9, at 41–42, U.N. Doc. A/3159 (1956). For views on the status of the exploitability test at the 1958 United Nations Conference on the Law of the Sea at Geneva, see 2 U.N. CONF. ON THE LAW OF THE SEA, GENEVA 1958, OFFICIAL RECORDS (PLENARY MEETINGS) 12–13, U.N. Doc. A/CONF. 13/38 (1958). *See also* U.N. Doc. A/CONF. 13/42, *supra* note 118, at 2 (Gros), 4 (Gutteridge), 9 (Alvarez Aybar), 19 (Whiteman), 24 (Buu-Kinh), 26–27 (Gomez Robledo), 31–32 (Paey), 34 (Carbajal), 40 (Whiteman), 42 (Nikolic, Wershof, Jhirad). But *see id.* at 5 (Rubio), 6 (Krispis), 10 (Calicede Castilla), 16 (Barros), 17 (Rosenne), 25 (Garcia Amador), 29–30 (Carty), 33 (Ruiz Moreno), 37 (Quarshie).

    On possible contraposed evaluations of the clause inserting the test, see W. Burke, Ocean Sciences, Technology and the Future International Law of the Sea 54–55 (Mershon National Security Program, Pamphlet Series No. 2, 1966).

148. For a discussion of criteria expanding the meaning of the exploitability test and disregarding the considerations which led to its adoption, see Goldie, *Contents, supra* note 93, at 11–14.

geographical and oceanological concept of the shoulder of the pedestal upon which the land masses rest—the geographers' notion of the part the continental shelf plays in the depiction of the world's oceanographic features. Indeed, this test is as free of any empirical connection with the ocean rim as are the much criticized claims of exclusive competence over vast areas of the oceans that Chile, Ecuador, Peru, and other Central and South American states assert (the "CEP claims"),[149] and as indeterminate—although for different reasons. The exploitability test's indeterminacy depends, no less than does that of the CEP claims, on factors not in the contemplation of the Law of the Sea Conference in 1958 when the participants agreed to its adoption and saw it as subordinate to the 200-meter test.[150]

Certain present-day developments, although still almost entirely in their experimental stages, add a new and heightened significance to the possibility that the exploitability test may, unless restrained, become a legal fiction permitting ever-increasing encroachments upon the inclusive uses of the high seas. For example, continuing recent improvements in direction drilling for oil and gas, Captain Jacques-Yves Cousteau's Conshelf I, II and III, the United States Navy's Sea-Lab, the development of new submersibles by the aerospace industry, Edwin Link's Man-in-Sea Project, and the discovery of a beckoning wealth of mineral nodules (as well as subsoil minerals, including petroleums) existing at great depths on and under the ocean floor,[151] all give the exploitability test a new importance—one promising to eclipse that of the 200-meter bathymetric contour line.

In his 1955 excoriation of the then developing legal doctrine of the continental shelf, Professor Georges Scelle pointed out[152] that, as then formulated by the International Law Commission, the doctrine could become little more than a fiction camouflaging the *"Faustrecht"*[153] of

149. For a discussion of the "CEP claims," see, *e.g.,* Goldie, *supra* note 87, at 31–38; Goldie, *supra* note 6, at 43, 49–51.
150. For a discussion of the juridicial significance of indeterminancy as used in this study, and especially with respect to the fictional qualities (which may be socially, as well as otherwise, beneficial) of indeterminate concepts, see J. STONE, LEGAL SYSTEM AND LAWYERS' REASONINGS 59, 263–67, 293, 299, 306, 317–20, 345–46 (1964); J. STONE, THE PROVINCE AND FUNCTION OF LAW 185–91 (1946).
151. For an impressionistic indication of man's increasing reliance on the sea as a source of food and resources and as an habitat, see Goldie, *The Management of Ocean Resources: Regimes for Structuring the Maritime Environment,* in 4 THE FUTURE OF THE INTERNATIONAL LEGAL ORDER, 155, 156–68 (C. Black & R. Falk eds. 1972).
152. Scelle, *Plateau continental et droit international,* 59 REVUE GÉNÉRALE DE DROIT INTERNATIONAL PUBLIC 5 (1955).
153. *Id.* at 19.

a series of ever-increasing claims into the common domain of the high seas, both upward to embrace the superjacent sea and the super-ambient air, and outward further into the oceans until a thalweg in the abyss is reached. He envisaged the free high seas and the great oceans as finally being enclosed within the territorial sovereignties of the coastal States. As it has thus far developed, however, despite the potential of the continental shelf doctrine for excusing unlimited territorial extensions into the high seas, an enclosure movement is not inevitable—provided that the definition of the continental shelf regions over which a coastal State may assert sovereign rights is made to rest exclusively on the test of depth, and the exploitability test is elim-inated. Because the exploitability test threatens to overthrow the 200-meter depth test and because it offers an ideal method for coastal States to augment extensively the submarine regions subject to their exclusive rights, it threatens to become one of the means for realizing Scelle's worst fears.[154] In overthrowing the uniform and certain, if for-malistic, definition of the continental shelf that the 200-meter bathymetric contour line provides, it opens up the possibility of a series of ever-increasing claims by States to exclusive rights to the oceans.

Alternative Meanings: A Problem of Definition and Scope

Merely to point to the potential extent of the exploitability test in terms of present and possible future scientific developments itself ex-poses ambiguities. The question arises whether Article 1 permits coastal States to assert sovereign rights to continental shelf areas be-yond the 200-meter bathymetric contour line if the resources of the zone claimed could be exploited by the application of the skills at the disposal of the world's technologically most advanced State. This position is opposed by authorities who argue that the test of ex-ploitability depends on the technological capabilities of the coastal State actually asserting a claim to exercise sovereign rights under the formula. For the purpose of the present discussion, the former construction of the exploitability test will be called the *absolute* stan-

---

154. The other is provided by the inclusion of "living organisms belonging to sedentary species" as among the "natural resources" of the continental shelf as provided in Article 2(4) of the Continental Shelf Convention. For a discussion of this potential threat to the freedom of the seas, see Goldie, *A Symposium on the Geneva Conventions and the Need for Future Modifications,* THE LAW OF THE SEA: OFF-SHORE BOUNDARIES AND ZONES 273, 285–90 (L. Alexander ed. 1967).

dard and the latter the *relative* standard.[155] Carl Franklin has expressed the former position with great clarity: "This depth which admits of exploitation should be interpreted *absolutely* in terms of the most advanced technology in the world, and not *relatively* in terms of the *particular* technology of any one coastal state."[156] This statement should be contrasted with the United States Senate Committee on Commerce's evaluation of the exploitability test in its Report[157] on the Marine Resources and Engineering Development Bill[158] (later the Act of 1966 of the same name).[159] The Committee reported: "[T]he Convention conveys both specific and immediate rights and prospective or potential rights, the latter to be acquired only as a result of national effort and achievement . . . ."[160]

Both versions of the exploitability test, the Senate Committee's no less than Franklin's, may be classified as "potential situation" interpretations of the exploitability test. Neither refers to the actual exploitability of a given continental shelf area measured by its resources or the exploitation actually being conducted or developed. Each of them, in its own terms, responds to a question based upon a hypothetical situation: what would be the sovereign rights of the coastal State in a claimed submarine region beyond its 200-meter isobath, (a) if that region were exploited by the world's most technologically advanced State, or (b) if the region in question were exploited at the technological level achieved by the coastal State itself? In neither situation is there any reference to what is being done, if anything is in fact being done, to exploit the submarine region being made the subject of a coastal State's claim.

155. This is to adopt from other branches of the law the dichotomy between relative and absolute tests. For the purposes of the present discussion of the exploitability test, the term *relative* will be taken as indicating that a coastal State will be deemed capable of acquiring sovereign rights only over those parts of a submarine region off coasts and beyond the 200-meter isobar that it can exploit on the basis of its own technological development, or that of enterprises licensed by it. The *absolute test,* on the other hand, permits the coastal State to treat as subject to its sovereign rights those submarine regions beyond the 200-meter isobar that could be exploited by the most advanced technologies. This latter test applies without regard to the technical capabilities of the coastal States or of the enterprises it licenses.

156. C. FRANKLIN, THE LAW OF THE SEA: SOME RECENT DEVELOPMENTS 23 (1961). Similar views have been expressed by Mouton and Young. *See* M. MOUTON, *supra* note 11, at 42, Young, *The Geneva Convention on the Continental Shelf: A First Impression,* 52 AM. J. INT'L L. 733, 735 (1958).

157. S. REP. No. 528, 89th Cong., 1st Sess. 11 (1965).

158. S. 944, 89th Cong., 1st Sess. (1965).

159. 33 U.S.C. §1101 (1970).

160. S. REP. No. 528, *supra* note 157, at 11.

In contrast to the interpretations just outlined, the International Law Commission treated the exploitability test as being relevant to existing situations of ongoing exploitations. For example, in its Commentary on Article 67 of its 1956 Articles Concerning the Law of the Sea, the Commission provided at least the starting point of a third position when it stated: "[T]he continental shelf might well include submarine areas lying at a depth of over 200 metres, but susceptible of exploitation by means of installations erected in neighboring areas where the depth does not exceed this limit."[161]

The submission here is that to give effect to the most literal interpretation of Article 1 of the Continental Shelf Convention and to remain consistent with the International Law Commission's thesis set out in 1956 (when it restored the exploitability test to its proposed definition of the continental shelf), the exploitability test should be viewed as allowing the coastal State no more than the power to assert its jurisdiction over exploitations beyond the shelf that began on the shelf, that is, those started on the landward side of the 200-meter isobath and subsequently extended beyond that line.[162] This construction would have the additional policy advantage of protecting the free high seas from the encroachments by coastal States that a less strict interpretation would permit and even encourage.

Separate Rights: No Possibility of Conflict?

A leading authority on the continental shelf, Professor F. V. Garcia Amador, has suggested that Article 1 contains two distinct sets of rights: those over the "continental shelf proper" and those derived from the exploitability test. He tells us that "[t]here is now a distinction between the broad categories of submarine areas and also, as in the conclusions of the Inter-American Conference, between two classes of rights: an existing right with regard to the continental shelf proper, and a potential right with respect to the other areas covered by the definition."[163] That is, rights over submarine areas depending

161. [1956] 2 Y.B. INT'L L. COMM'N 296, 11 U.N. GAOR Supp. 9, at 41, U.N. Doc. A/3159 (1956).

162. That this may not be the actual outcome of interpretation but that the test under discussion may provide a vehicle (albeit spurious but nevertheless strongly emphasized) for the promulgation of claims to the seabed and subsoil in addition to the volume of the waters and the superambient air, is the topic of Goldie, *supra* note 154, at 285–90.

163. F. GARCIA AMADOR, THE EXPLORATION AND CONSERVATION OF THE SEA 111–12 (1959). The potential right of which Garcia Amador writes may, perhaps, be

on the exploitability test are to be viewed as "potential rights." They appear, at first blush, to be rights that spring up in favor of the coastal State, without any previous and contraposed right having been vested in some other State or citizen thereof. That this is in fact the case will be contested, and the theory suggesting that rights created by the exploitability test are no more than potential rights will be questioned. In creating new exclusive rights, the test must inevitably have a divesting effect upon preexisting general international law rights. For example, the general international law right of exploring a submarine region is terminated when the area becomes exploitable. This may be viewed as being substantially analogous to the common law concepts of a fee simple determinable, a fee simple subject to an executory interest, or (notwithstanding the fact that the continental shelf region is said to vest *ipso jure* in the coastal State) a power of termination.[164] This last analogy is all the more apt because, although the framers of the continental shelf doctrine envisaged it as automatically conferring sovereign rights on the coastal States without the need for the performance of dispository or occupying acts, the exploitability test remains inevitably and essentially subjective in its application. Hence, the theory that the exploitability test creates potential rights also involves the exposure of existing rights to extinction. Thus, Garcia Amador's presentation contains a suppressed catagory of "potential exposures."

The formulation of the exploitability test, "to where the depth of the superjacent waters admits of the exploitation of the said areas," does not include any reference to exploration. This formulation need not be viewed, and, it is submitted, should not be viewed, as including exploration by a necessary implication. For, whereas Article 1 merely speaks of "exploitation" without mentioning exploration, Articles 2, 4, and 5 speak of "exploration" and "exploitation" in tandem. This surely indicates that the full range of activities over which the coastal State is entitled to exercise sovereign rights is more extensive than the range of the exploitability test operating as a condition. An elucidation

---

thought of as resembling, at least in its temporal and ex post facto features, a condition subsequent in the Anglo-American common law—indeed it presents us with a conditional right which upon actualization defeats existing rights; frequently it will even defeat the existing rights of those who brought about its operative facts.

164. Feith, *supra* note 93, at 125, 127, 131–32; [1953] 2 Y.B. Int'l L. Comm'n 214, 8 U.N. GAOR Supp. 9, at 14, U.N. Doc. A/2456 at 15, para. 72 (1953). In its 1953 Draft Report [U.N. Doc. A/CN. 4/L. 45/Add. 1 (1953)], the International Law Commission stated: "The rights of the coastal State over the continental shelf are independent of occupation, actual or fictitious. They belong to it *ipso jure*." *See*, however, comments made by the Commission's members of this formula, [1953] 1 Y.B. Int'l L. Comm'n 347–50.

of this discrepancy between Article 1 and Articles 2, 4, and 5 requires that a preliminary point should be made, then a question put.

The preliminary point is this: since different resources call for different techniques for their exploitation, the exploitability test, in order to remain true to its meaning, should be applied as a basis for extending a coastal State's sovereign rights over the exploitation of only such specific resources as are at that time exploitable. For example, although manganese nodules may be won from great depths in the near future,[165] any concept of exploitability based on that potential would be irrelevant to mining for solid minerals in the subsoil of a submarine continental terrace. Similarly, an application of the exploitability test, which might well be relevant to taking oil and gas, would be irrelevant to a claim to exercise exclusive continental shelf rights over a sedentary fishery.[166]

Now the question may be asked. If exploration is not a part of the exploitability test, then a coastal State may not view exploration activities beyond its adjacent continental shelf (as determined by the criterion of depth—assuming the zone of the activities in question to be beyond the territorial sea) as falling under the sovereign rights conferred upon it by the continental shelf doctrine. But if the contents of a submarine area's subsoil remain unknown, how can the exploitability test apply in cases other than those of continuing an exploitation out from the 200-meter line and beyond the territorial sea? The advocates of a more extended scope for the exploitability test may be tempted to answer that a coastal State has, in common with all States, a right to regulate its own citizens' exploration activities, no less than their exploitation activities, beyond the continental shelf adjacent to its coasts and to conduct such exploration itself. This being so, the coastal State may attach its sovereign rights, by virtue of the exploitability test, to exploitations begun under its aegis and beyond its shelf resulting from explorations conducted by its citizens (or by corporate entities established under its laws) under the general

165. *See, e.g.*, J. MERO, THE MINERAL RESOURCES OF THE SEA 127, 277–79 (1965).
166. The whole question of the "quantum" (to borrow yet again from the common law of real property) of an interest that the exploitability text purports to vest in the coastal State has not been faced by writers in international law. Yet this question, in whatever terms and on whatever analogies it is presented, is necessarily central to any thesis arguing for a liberal interpretation of the exploitability test. This writer has argued elsewhere that the economic interests coastal States purport to protect by extending their jurisdictions seaward under the banner of the exploitability test would be better protected by other institutional means, *e.g.*, the treaty regime proposed by this writer. *See* Goldie, *supra* note 154, at 273–95; Goldie, *Contents, supra* note 93, at 38–53.

international law privileges that permit all the citizens of all the States of the world to explore for and exploit the resources of the oceans beyond the jurisdictional areas of States on the ground that the high seas are open to all.

There is a central difficulty to this position. If one coastal State claims to exercise authority over activities in a submarine area on the premise of what general international law, rather than the Continental Shelf Doctrine, allows, then so may any other State. That is, any noncoastal State or its citizens may, under general international law and based on this reasoning, *explore* for resources beyond the coastal State's territorial waters and 200-meter isobath but must do so in its immediate vicinity, in full equality with the coastal State's citizens. Although a noncoastal State's citizens, operating under their country's laws, may thus discover a resource outside, but in the general vicinity of, the 200-meter bathymetric contour line (or, perhaps, territorial sea) surrounding the coastal State, that noncoastal State is precluded, by the liberal construction of the exploitability test, from exercising the exclusive authority over the resource that general international law would allow. The potential rights that the exploitability test vests in the coastal State operate to divest the very States under whose laws the exploration activities were initiated and carried through to the point of establishing the possibility of an economic return. Again, *explorations* begun on the landward side of the 200-meter isobath, and continued beyond it, may not validly be viewed as extending the exploitability test's protection to exploitations that are begun beyond that isobath and have no starting point on its landward side. Finally, as has been previously observed, the exercise of sovereign rights beyond the 200-meter isobath over the exploitation of one resource (being an exploitation activity begun on the landward side of that bathymetric contour line) would not justify coastal States' claims to exercise any similar authority over the exploitation of different resources beyond the 200-meter isobath that had begun there, rather than within that contour line. For example, the exercise of sovereign rights over activities beyond the 200-meter line exploiting oil and gas resources would not, of itself, justify a claim to exercise similar rights over the exploitation of sulphur deposits.

Thus, because the term *exploration* is not expressly joined to *exploitation* in Article 1's formulation of coastal States' rights over adjacent continental shelf regions, there would appear to be little justification for treating rights defined only in terms of exploitation as necessarily including rights defined in terms of exploration. But if the exploitability test is referable only to exploitations beyond the 200-

meter bathymetric contour line, then, because the resource to which the test is to be applied must be known before it can be said to be exploitable, the test can only relate to exploitation activities begun on the landward side of that line. The distinction between exploration beyond the continental shelf under the privilege subsumed under the doctrine of the freedom of the high seas and exploration conducted on the shelf by virtue of the continental shelf doctrine and the Convention's first three Articles, which embody that customary international law doctrine, should not be overlooked. (Clearly, an exploration beyond what are then the outer limits of the continental shelf cannot be said to be exploration conducted under either the Convention or the customary law continental shelf doctrine.) Furthermore, the burden of proof is upon those who argue for the test's extension to say that as soon as an exploration shows a resource to be exploitable, at that moment the coastal State's sovereign rights under the continental shelf doctrine may leapfrog out to a new and deeper bathymetric contour line. Finally, this analysis points to this anomaly: a noncoastal State may explore for seabed resources beyond the shelf, but the coastal State may immediately claim the fruits of the exploration if that noncoastal State's activity invests those resources, under the interpretation of the exploitability tests now being appraised, with the status of becoming resources of the shelf by virtue of that exploration. This kind of contingent interest, which, upon their discovery, immediately shifts the rights in the resources from the State claiming them under the freedom of the high seas (and whose conduct prompted that shift) to the passively waiting coastal State, would appear to be inconsistent with both reason and stability of expectations.

To summarize the preceding discussion: the supporters of the exploitability test's liberal extension would argue that, although all States and their citizens enjoy privileges of *exploring* the oceans' resources beyond the jurisdiction of the coastal State, whenever an exploration shows that a submarine resource adjacent to an existing zone of continental shelf jurisdiction is exploitable, then that zone is transformed from the category of "high seas" to that of "continental shelf." Simple as such a solution might appear, it is fraught with difficulties. For if Article 1 should be viewed as changing the character of submarine areas beyond the continental shelf (however defined) from high seas areas open to exploration activities conducted under the laws of any and all States, distant or coastal, to areas subject to the coastal State's exclusive sovereign rights over exploration and exploitation activities, then the Article should set out the terms of such a divesting condition's operation and give some guidelines regarding the treatment of the equities involved. In addition, some rules governing possible conflicts of claims and governing laws should, in time, be provided—perhaps by a "critical date" theory.[167]

The fact that the Convention's framers never provided safeguards protecting the equities that the divesting effect of the exploitability test would place in jeopardy surely indicates that in their view Article I did not contemplate creating these shifting rights. To argue now for the recognition of such rights as inhering in the exploitability test points to a further paradox. To apply that test in the contingent manner envisaged, as for example in Garcia Amador's exposition, would be to divest the enterprises whose explorations first established the area's exploitability and as a result of whose efforts the coastal State's continental shelf claims became possible. Instead of being applied to the advantage of those who made the exploitability test concretely applicable to a given situation, it would operate to deprive them of the fruits of their labors and transfer those fruits to the governance and taxation of a State that had merely remained a passive onlooker—for it, too, might have acted on the same basis as the State whose citizens had invoked the privileges of the freedom of the seas. Furthermore, what would be the result of legislation by the coastal State requiring that all continental shelf exploitations be carried on by its citizens or by corporations created under its laws and in which either the State or its citizens had a controlling interest? Thus it becomes clear that Article 1 is not adequate to support arguments justifying liberal extensions of the submarine areas subject to coastal States' continental shelf sovereign rights.[168] Asserting the contrary position amounts to treating the Article as a divesting clause, as already indicated, and to according coastal States unlimited discretion to determine unilaterally what submarine regions are to fall, under international law, within their exclusive sovereign rights. Such a power would make the freedom of the high seas a merely determinable interest at a coastal State's election to exercise a right of entry into the mineral estate of the original owner. It is submitted that this result should be scrupulously avoided, for although international law permits States to exercise unexaminable discretions in many areas of international relations, one of the major tasks of contemporary international law is to reduce and tame the scope of those discretions, not add to them.

As a practical matter, the recognition of such widely extending

---

167. For a discussion of this concept, see Goldie, *The Critical Date*, 12 INT'L & COMP. L.Q. 1251 (1963).

168. For an example of such arguments, see note 155 *supra and accompanying text.*

potential rights as these to be enjoyed by coastal States would have devastating repercussions. From the analogy, raised in the preceding paragraphs, of such common law concepts as a fee simple determinable or a fee simple subject to an executory interest, it can be asked, would not the assertion of a potential right (upon the fulfillment of the conditions for that right's becoming an "actual right") create opportunities, and indeed almost irresistible pressures, for power confrontations between the subsequently claiming coastal State and the noncoastal State, whose citizens' lawful exploratory activities made the coastal State's divesting claim possible in the first place? And would not such confrontations generally lead to the deterioration of the relations between those States down to the level of the old *Faustrecht* that has plagued international relations for so long?

In addition to this criticism of the outcomes and values of Professor Garcia Amador's argument, it is necessary to emphasize that, from the point of view of States and individuals whose rights are divested by the operation of the exploitability test, a distinction between coastal State claims in terms of depth and coastal State claims in terms of exploitability will have little relevance; the test generates one ever-encroaching form of exclusive claim. Rights formerly derivable from the freedom of the seas become divested as submarine areas under the regime of the continental shelf. Garcia Amador's distinction between the concepts of depth and exploitability thus boils down to asserting that coastal States may, under appropriate circumstances, have alternative means of bringing submarine zones within their exclusive governance—when the background of the International Law Commission's selection of the term *adjacent* in its 1956 Article 67 defining the continental shelf is borne in mind.[169] Once a zone has been brought within the continental shelf regime, it becomes subject to a coastal State's claim of exclusivity. This does not differentiate between depth and exploitability when confronting noncoastal States' inclusive claims advanced under the doctrine of the freedom of the seas.

169. *See* [1956] 1 Y.B. Int'l L. Comm'n 1, 127, 130–41 (especially François, at 130); [1956] 2 Y.B. Int'l L. Comm'n 297, 11 U.N. GAOR Supp. 9, at 42, para. 8, U.N. Doc. A/3159 (1956). The term *adjacent* was accepted without question by the Fourth Committee, the Drafting Committee and the Plenary Meetings of the 1958 Conference on the Law of the Sea at Geneva. *See* 2 U.N. Conf. on the Law of the Sea, Geneva 1958, Official Records (Plenary Meetings) 11–13, 89, 91, 93, U.N. Doc. A/CONF. 13/38 (1958); U.N. Doc. A/CONF. 13/42, *supra* note 118, at 31–45, and note especially the Panama proposal, which used the term *adjacent* with respect to the total continental terrace region, U.N. Doc. A/CONF. 13/C. 4/L. 4 (10 Mar. 1958), *id.* at 127.

*Counterpoint of Interests—Freedom of the High Seas and
Encroachment of Exclusive Claims*

As early as May 1966, the Commission to Study the Organization of
Peace recommended that no nation should be allowed to extend the
limits of its national jurisdiction beyond twelve miles for fishing or the
200-meter bathymetric contour line or "some other readily definable
boundary" for continental shelf activities.[170] In March 1969 the Commission clarified its position by recommending: "The Convention on
the Continental Shelf should be revised so as to provide that national
exploitation rights over sea-bed resources end at the 200-meter depth
line or 50 nautical miles from shore, whichever occurs further out."[171]
As an intermediate measure, it recommended that States should "re-
frain from granting leases to mineral resources beyond the 200-meter
depth limit."[172]

In the meantime, opposition developed both to this thesis and to any
amendment of the Continental Shelf Convention. For example, in July
1968 the Committee on Deep Sea Mineral Resources of the International Law Association's American Branch offered, in its Interim
Report, the following conclusion:

Since exploration and exploitation of undersea minerals is likely to occur
earlier in the shallower waters of the oceans adjacent to the continents than in
the abyssal depths, it follows that if jurisdictional uncertainties arise to
impede such operations during the next several decades, such problems will be
primarily related to the scope of the mineral jurisdiction which is already
vested exclusively in the coastal states by the "exploitability" and "adjacency" criteria of jurisdiction which now appear in the Continental Shelf
Convention. This uncertainty, if necessity for its resolution occurs, might be
removed by consultation among the major coastal nations which are capable
of conducting deep sea mineral development, looking toward the issuance by
those states of parallel ex parte declarations. These declarations might appropriately restrict claims of exclusive sea-bed mineral jurisdiction, pursuant to
the exploitability and adjacency factors of the Continental Shelf Convention,

170. Commission to Study the Organization of Peace, New Dimensions for the United
     Nations: The Problems of the Next Decade 45 (1966).
171. Commission to Study the Organization of Peace, The United Nations and the Bed
     of the Sea 24 (1969). Senator Pell has proposed, in Declaration of Legal Principles
     Governing the Activities of States in the Exploration and Exploitation of Ocean
     Space, S. Res. 33, 91st Cong., 1st Sess., art. VI (1969), that the outer limit of the
     continental shelf be set at the 550-meter isobath or a distance of 50 miles from ter-
     ritorial sea base lines, whichever results in the greater continental shelf area. *See
     Hearings on S. Res. 33 Before the Subcomm. on Ocean Space of the Senate
     Comm. on Foreign Relations,* 91st Cong., 1st Sess., at 9, 14 (1969).
172. *Id.* at 25.

to (i) the submerged portions of the continental land mass, limiting this provisionally to a depth of, say, 2,500 meters, or (ii) to a stated distance (say 100 miles) from the base line, whichever limitation encompasses the larger area. Such declarations might appropriately recognize special cases. Two such classifications suggest themselves:

(i) In the case of states whose coasts plunge precipitously to the ocean floor (e.g., on the west coast of South America), the suggested 100-mile limit on sea-bed mineral jurisdiction would automatically operate on the deep ocean floor.

(ii) In the case of narrow or enclosed seas, the principle of adjacency might appropriately carry coastal mineral jurisdiction to the median lines, even though these are beyond the continental blocks.[173]

This approach presaged the line that the great majority of the supporters of a broad shelf were to follow, namely, a complicated and questionable exegesis from the Continental Shelf Convention, based on a selective review of its *travaux préparatoires*. An alternative would have been the advocacy, on straightforward policy grounds, of the need to recognize now, no matter what the position might have been in the 1950s, that technology and investment call for a wide shelf. Be that as it may, in 1969 the National Petroleum Council followed up the American Branch Committee's statement with its own publication, *Petroleum Resources Under the Ocean Floor,*[174] further developing the position of the American Branch Committee but leaving aside the earlier publication's denomination of either a bathymetric contour line or line of distance (whichever should encompass the larger area) as a "provisional" delimitation of the shelf's outer limits. The National Petroleum Council proposed:

[S]ince the plunge of the slope has often been locally overlapped extensively by the sediments of the continental rise, a boundary just oceanward of the base of the slope, to include the shelf, the slope and the landward portion of the continental rise, where developed, more closely approaches the true ocean-bottom boundary between continental and oceanic areas and is the most natural and appropriate outward limit of a country's sovereign rights over bottom resources. A boundary thus drawn gives recognition to the natural oceanward extension of the domain of each coastal nation and the inclusion under its jurisdiction of that suboceanic territory over whose natural resources the coastal nation is most practically suited to exercise control.

In summary, given a recognition of the above scientific facts, it is apparent that the outer edge of the continent is a far more logical choice than the outer

173. Comm. on Deep Sea Mineral Resources of the Am. Branch of the Int'l L. Ass'n, Interim Report, July 19, 1968, at XVII–XVIII [hereinafter cited as Am. Branch Comm. Interim Rep.].
174. NPC Report, *supra* note 96, at 55–67.

edge of the geological continental shelf as the limit of coastal-nation exclusive
jurisdiction over the resources of the seabed and subsoil. The participating nations at Ciudad Trujillo in 1956 and at Geneva in 1958 wisely declined to limit
the coastal nation's exclusive jurisdiction to the geological continental shelf or
to the 200-meter isobath.[175]

Although the National Petroleum Council has undertaken to represent American business and common sense, its thesis has not gone
uncriticized by industrial interests other than oil. For example,
Malcolm R. Wilkey, general counsel for the Kennecott Copper Corporation, has been outspoken in his dissent from the Council's
position.[176] The present writer strongly suggests, furthermore, that
the criticism he leveled at the American Branch Committee's Report
during the Deep-Sea Mining Session[177] of the International Law
Association's Fifty-Third Biennial Conference at Buenos Aires on
August 25-31, 1968, applies *mutatis mutandis* to the relevant pages of
the National Petroleum Council's Report. Whereas the deep-ocean
sciences and engineering technologies advance and markets fluctuate
in response to production and need, the exploitability test must always
have a contingent operation. Even the probability of a brilliant future
for the human exploitation of the resources of the ocean bed and
subsoil cannot remove the uncertain qualities of time and utilization.
Nonetheless, the National Petroleum Council Report entirely ignores
the contingent quality of the exploitability test. In contrast to the
clear words of Article 1 of the Continental Shelf Convention, the language of the National Petroleum Council's interpretation of the exploitability test is that of a sanguine beneficiary who might ignore intermediate interests and regard a contingent gift in his favor as though
it had immediately vested in him. Perhaps the major misfortune of the
National Petroleum Council's argument is that, by its attempts at
legerdemain in the name of legislative history, it repels at least some
of the members of the scientific-educational estate, who might otherwise be persuaded of the feasibility of the broad continental shelf
policy by a full and frank discussion of blueprints for national policy
from the point of view of the oil industry's perception of its own needs
and those of the country and the world. This argument is reinforced
by the current debates in the United Nations General Assembly and
its Seabeds Committee indicating that the broad continental shelf is
having a far stronger appeal for the developing countries than the narrow shelf thesis. This, in its turn, is currently assisting to jeopardize

175. *Id.* at 67.
176. M. Wilkey, The Role of Private Industry in the Deep Ocean 17-19 (Symposium on
    Private Investments Abroad, Southwestern Legal Foundation, 1969).
177. 53 INT'L L. ASS'N REP. 200, 203-06 (1968) (Goldie).

the chances of an effective and well-endowed regime for the deep-ocean floor, since the focus of attention is tending to shift from the regime to individual States' more extended claims over offshore regions as future sources of developmental wealth.

The National Petroleum Council's advocacy of a broad continental shelf was roundly criticized by the Commission on Marine Science, Engineering and Resources (the Stratton Commission), which concluded its reasons for rejecting the proposal as follows: "The NPC proposal is also unfair to the inland nations of the world. . . . U.S. action to effectuate the NPC proposal would be regarded as a 'grab,' even if all the coastal nations followed suit."[178]

Finally, the National Petroleum Council's definition of the continental shelf by reference to its own interpretation of the exploitability test[179] leaves out of account the critical fact that the subject of its definition of the continental shelf is: the continental pedestal constituted by the continental dry land + continental shelf + continental slopes + continental slopes' toe + continental rise and/or terraces. In seeking to bring these other submarine features within the term *continental shelf,* the National Petroleum Council is committing the elementary solecism of trying to force a part to include the whole.

A Preliminary Step—Eliminate the Exploitability Test

The Advantages

The thrust of the preceding argument has been to show that when the exploitability test was incorporated into Article 1, it was not intended

178. COMM'N ON MARINE SCIENCE, ENGINEERING & RESOURCES, OUR NATION AND THE SEA 145 (1969) [hereinafter cited as STRATTON COMM'N REPORT]. The Stratton Commission was appointed by President Johnson on Jan. 9, 1967 pursuant to the Marine Resources and Engineering Development Act of 1966, 33 U.S.C. §1101 (1970). The Report was accompanied by the reports of eight panels bound into three volumes: 1 SCIENCE AND ENVIRONMENT, 2 INDUSTRY AND TECHNOLOGY: KEYS TO OCEAN DEVELOPMENT, 3 MARINE RESOURCES AND LEGAL–POLITICAL ARRANGEMENTS FOR THEIR DEVELOPMENT (1969) [hereinafter cited as STRATTON COMM'N PANEL REPORTS and prefixed by the appropriate volume number]. The roman numeral prefixing the page numbers in citations to the Panel Reports indicate the Panel cited. In commenting on the NPC Report, the International Panel note 173 *supra,* states: "There is little question but that the NPC view of adjacency extends too-far beyond the 200 meters, the depth of most geological shelves of the world. Considering the totality of its interests in the oceans, the United States would never have accepted the Convention of the Continental Shelf as NPC now reads it." 3 STRATTON COMM'N PANEL REPORTS VIII–20.

179. *See* text accompanying notes 173–174 *supra.*

to compete with the continental shelf's definition in terms of the 200-meter isobath but rather to serve in an auxiliary and subordinate capacity. This paragraph is, however, completely *de lege ferenda*. Since 1958, scientific and technological developments have radically altered the test's potential operation, so that now it threatens to render the 200-meter test obsolete. Whereas the 200-meter bathymetric contour line was chosen as a test because it clearly and unequivocally indicated the outer limits of the submarine areas within which States could exercise their rights, the exploitability test, largely because of its indeterminacy, permits the assumption of sovereign rights without regard to depth.[180] It therefore may be available (on mistaken premises, it is true) to excuse virtually unlimited extensions of rights beyond the limiting depth of 200 meters vertically and beyond the land-encircling–200-meter bathymetric contour line horizontally. Hence, this formula, which in 1958 was thought to have a subordinate function, promises to supplant in the next decade the definition to which it had been originally attached as an auxiliary and supplementary concept. The greater irony, however, is that the very test chosen for its fixity and certainty in preference to the indeterminacy of the break-in-slope criterion now is in peril of being superseded by a test of even greater indeterminacy than the one originally rejected in its favor.

Because the present-day technological and economic developments already indicated seem irreversible,[181] and because the current trend of ocean science and technology points to man's ever-increasing conquest of the oceans and their seabed and subsoil, the exploitability test promises to promote and multiply the instances of encroachment on the freedom of the seas.[182] It thus promises to become an increasingly

---

180. In a similar vein, without, however, asserting the possibilities just indicated, Joyce Gutteridge, assistant legal adviser of the United Kingdom Foreign Office (and a participant in the Fourth Continental Shelf Committee of the 1958 United Nations Conference on the Law of the Sea) has written: "The present definition is bound to result in uncertainty; and may lead to disputes between States in cases where the same continental shelf is adjacent to the territories of opposite or adjacent States, or, at the least, to difficulties in fixing by agreement the boundaries of such shelves. Moreover, 'exploitability' is a subjective criterion. It may well be asked, as it was asked at the Conference, how it is to be determined that a particular submarine area beyond the depth of 200 metres 'admits of exploitation.' " Gutteridge, *supra* note 138, at 110. Miss Gutteridge, with Admiral Mouton, sponsored a joint Netherlands–United Kingdom proposal to eliminate the exploitability test but extend the depth criterion to the 550–meter isobath. *See* Netherlands and Great Britain: Proposal, U.N. Doc. A/CONF. 13/C. 4/L.32, in U.N. Doc. A/CONF. 13/42, *supra* note 118, at 135. Generally, for the jurisprudential significance of indeterminacy, see J. STONE, LEGAL, *supra* note 150, at 59, 263–67, 293, 299, 306, 317–20, 345–46; J. STONE, PROVINCE, *supra* note 150, at 185–91.
181. *See* Goldie, *supra* note 151, at 156–68.
182. *See* Goldie, *supra* note 6, at 56–59.

dangerous threat to the peaceful and equitable allocation of the oceans' mineral resources and therefore should be eliminated from the legal definition of the continental shelf. Not only should it be deleted from Article I of the Continental Shelf Convention, but it should also be excluded from any doctrines that may be developed for interpreting and applying that Convention.

The Problem of Vested Rights

Although this discussion points to the pressing need for important modifications of Article 1 in order to equip the exploitability test with modalities that would ensure its reasonable application, it even more strongly indicates the advantages of deleting that test altogether. In the long run, the deleterious effects on world community goals of unilateral claims of exclusivity made possible by the exploitability test will far outweigh any short-run benefits it may have. There should, however, be a savings clause to preserve established rights (at least analogous to property rights on dry land) stemming from grants by coastal States to explore for and/or exploit the mineral resources of submarine areas extending beyond the 200-meter or the 100-fathom isobath.[183] Merely to delete the exploitability test could, without some savings clause, be treated as a basis for making pointless expropriations—since the basis of some rights could be seen as having been taken away by virtue of that deletion. If, despite the foregoing arguments, however, the exploitability test should be retained as one of the criteria for allocating continental shelf rights beyond the 200-meter isobath, then the exploitability test should be linked with pollution control criteria.

The Continental Shelf and Pollution Prevention: Competences and Duties

Article 24 of the Convention on the High Seas calls upon States to prevent marine pollution by oil, and Article 25 obliges them similarly with respect to dumping radioactive materials into the sea. Accordingly, these provisions should be read with Article 5, paragraphs 1

183. For an indication (by way of example) of the extent of the interests operating off the shores of the United States (and not including similar operations off the shores of other countries) that could be threatened by expropriation if effective protective measures of rights that had become vested were not taken, see Goldie, *Exploitability Test, supra* note 94, at 452–55, 462–76.

and 7, of its companion Continental Shelf Convention in order to plot the limits of the obligations of States to prevent or at least to regulate pollution in their shelf areas. Article 24 of the Convention on the High Seas provides: "Every State shall draw up regulations to prevent pollution of the seas by the discharge of oil from ships or pipelines or resulting from the exploitation and exploration of the seabed and its subsoil, taking account of existing treaty provisions on the subject." And Article 25 prescribes:

Every State shall take measures to prevent pollution of the seas from the dumping of radioactive waste, taking into account any standards and regulations which may be formulated by the competent international organizations.

All States shall co-operate with the competent international organizations in taking measures for the prevention of pollution of the seas or air space above, resulting from any activities with radioactive materials or other harmful agents.

Article 5, paragraph 1, of the Continental Shelf Convention asserts: "The exploration of the continental shelf and the exploitation of its natural resources must not result in any unjustifiable interference with navigation, fishing or the conservation of the living resources of the sea, nor result in any interference with fundamental oceanographic or other scientific research carried out with the intention of open publication." And, more specifically, paragraph 7 provides: "The coastal State is obliged to undertake, in the safety zones, all appropriate measures for the protection of the living resources of the sea from harmful agents."

It is of interest to note that neither in the writings of publicists nor in the deliberations of the International Law Commission and other United Nations agencies, commissions, committees, working groups, and secretariats have there been proposals that these provisions should be read with the exploitability test in Article 1 of the Continental Shelf Convention so as to limit the concept of exploitability to what is manageable.

The Santa Barbara Channel disaster of 1969 underlines for us all that it is easier to drill a submarine oil well than to cap it after a blowout.[184] If newspaper reports of the fire and blowout at the Chevron Oil Company's well near Venica, Louisiana, are any indication, the lessons of Santa Barbara have not yet been learned.[185] Commenting on Senator Pell's Senate Resolution 33 of 1969,[186] the

184. *See, e.g.,* N.Y. Times, Jan. 31 to Apr. 3 1969 *passim.*
185. *See* N.Y. Times, Mar. 2, 1970, at 17, cols. 1–6.
186. S. Res. 33, 91st Cong., 1st Sess., art. VI (1969), recommends that the President place a resolution endorsing basic principles for governing the activities of nations

present writer suggested: "Senate Resolution 33 [should] contain a pledge that no exploration of exploitation activities will be espoused or licensed by states, or by any international organizations, at depths greater than the feasibility of closing of blowouts. Nor should pipelines be permitted below . . . depths [at which they may be rapidly repaired]."[187]

The pledge referred to in this quotation is, of course, a promise by States parties to the Declaration of Legal Principles, which Senator Pell included in his Resolution, that they promulgate the necessary domestic legislation to prohibit drilling wells and laying pipelines below the depths of rapid and complete repair. Indeed, while exploitability remains a test for determining the outer limits of the continental shelf, the technological capacity to control the consequences of drilling holes in the seabed, rather than the mere capability of promiscuously inflicting them on the environment, should set both the outer limit of exploitations and of the meaning of exploitability as a criterion of the extent of coastal States' continental shelves under Article 2 of the Continental Shelf Convention.

Intermediate or Trusteeship Zone Proposals

A Review of Some Proposals

The idea that a zone of the seabed and subsoil lying between the 200-meter isobath and the continental rise (the area that, if beyond the territorial sea, the "wide shelfers" would subsume under the coastal State's continental shelf and "narrow shelfers" would attribute to an international regime) should be brought under some kind of a compromise regime has been formulated in terms of several different proposals. Those of most interest are Professor Henkin's "buffer zone,"[188] the Stratton Commission's "intermediate zone"[189] and the "trusteeship area" recommended in the 1970 United States Working Paper Draft United Nations Convention on the International Seabed Area (United States Draft Convention).[190]

---

in ocean space before the United Nations Committee on the Peaceful Uses of the Seabed and Ocean Floor beyond the Limits of National Jurisdiction. *See* note 171 *supra.*

187. *Hearings, supra* note 170, at 290, 300 (memorandum of L. F. E. Goldie).
188. L. HENKIN, *supra* note 96, at 46–48.
189. STRATTON COMM'N REPORT 151–53.
190. United States Working Paper "Draft United Nations Convention on the International Seabed Area," *tabled* Aug. 3, 1970, United Nations Gen. Assembly Comm.

These concepts will be discussed on their merits and without reference to the legal difficulties inherent in establishing them as universally applicable institutions. For example, one problem is determining whether the convention that constitutes an international or quasi-international seabed area seaward of the 200-meter isobath and within a coastal State's adjacent continental shelf by virtue of its exploitability is merely contractual or whether it would amount to an instrument of permanent dedication by the parties and so remain international after a State had withdrawn from the treaty.[191] Secondly, the first two of the above proposals do not deal with the problem of the means of the defeasance of the submarine area seaward of the 200-meter bathymetric contour line but within the coastal State's adjacent continental shelf by virtue of its exploitability. While the third proposal does deal with the issue, its recommendation that such sovereign rights should not be recognized by the parties to the proposed Convention, despite their lawfulness, clearly could give rise to conflicts of will and even force.[192]

The Buffer Zone

Professor Henkin's buffer zone would be within the regime of the "deep sea for all purposes."[193] Accordingly, claims of sovereign rights for coastal States may be expressly denied. Thus, he sees the buffer zone as falling within the regime of the deep ocean floor subject, however, to allowing the coastal State alone to have the right of exploiting the resources of the area.[194] He argues that this would give the proponents of the broad continental shelf substantially what they are asking for. Secondly, vested rights in exploitations beyond the 200-meter isobath would be protected from the possibility of defeasance[195] should they pass from the coastal State's regime of the continental

---

on the Peaceful Uses of the Seabed and the Ocean Floor beyond Nat'l Jurisdiction, U.N. Doc. A/AC. 138/25. *See* Committee on the Peaceful Uses of the Seabed and Ocean Floor beyond the Limits of Nat'l Jurisdiction, Report, 25 U.N. GAOR Supp. 21, at 130, U.N. Doc. A/8021 (1970).

191. For an analysis of this and cognate difficulties raised and not resolved by the United States Draft Convention, see Jennings, *United States Draft Treaty on the International Seabed Area—Basic Principles,* 20 INT'L & COMP. L.Q. 433 (1971).

192. *See, e.g.,* Jennings, *supra* note 191, at 444, 446–47.

193. L. HENKIN, *supra* note 96, at 47.

194. *Id.*

195. On the problem of the defeasance of vested rights in exploitations beyond the 200-meter isobath, see *supra* at 52–55.

shelf to the international regime of the deep oceans.[196] Henkin suggests that this principle could be incorporated in a "revision of the [Continental] Shelf Convention (or in a protocol to it), even before a law of the deep sea is developed."[197] Another preparatory proposal he makes is that if the suggestion of the buffer zone commends itself, then the "United States could further it diplomatically or by unilateral declarations."[198] He offers an example of this unilateral promotion: "[T]he United States might declare its commitment to the 200 meter definition and express its willingness to submit its interim exploitations beyond that depth to whatever regime for the seas is ultimately agreed. At the same time it would oppose any foreign mining installations within X miles from its shores and respect a similar zone established by other coastal states."[199]

The Intermediate Zone

The Stratton Commission proposed that "intermediate zones be created encompassing the bed and subsoil of the deep seas, but only to the 2,500-meter isobath, or 100 nautical miles from the baseline for measuring the breadth of each coastal nation's territorial sea, whichever alternative gives the coastal nation the greater area for the purposes for which intermediate zones are created."[200]

Some reservations should first be mentioned with regard to the Commission's selection of the 2,500-meter isobath as the outer limit of the intermediate zone. The concept of the intermediate zone has been offered by the supporters of the narrow shelf as a mediating position between their view and that of the wide-shelf advocates. That isobath, we may presume, is put forward to indicate the outer limits of the broad international shelf as representing the true geological or topographical boundary on the sea floor between the crust of the deep-ocean basins and the continental landmasses. This geological boundary is said to exist as an empirical fact. One may be confident,

196. L. HENKIN, MINERAL RESOURCES, *supra* note 96, at 47.
197. *Id.*
198. *Id.* at 47–48.
199. *Id.* at 48, n. 149.
200. STRATTON COMM'N REPORT 151. *See also* 3 STRATTON COMM'N PANEL REPORTS VIII–34–45, where the following recommendation was offered: "[I]t is recommended that the outer limit of the intermediate zone be defined in terms of the 2,500 meter isobath or 100 nautical miles from the base lines for measuring the breadth of the territorial sea, whichever alternative gives the coastal State the great submarine area for the purposes for which the intermediate zone was credited."

therefore, that it does not exist in all places exactly at the 2,500-meter isobath.

For example, the Committee on Deep Sea Mineral Resources of the American Branch of the International Law Association, in its Interim Report of 1968, based the thesis it then entertained—that the outer limits of the continental shelf (subject to coastal States' exclusive jurisdiction) should be at the 2,500-meter isobath—on the work of Dr. William T. Pecora of the United States Geological Survey. The Committee argued that there is so "marked [a] change of structure between the continental mass and the crust of the deep ocean basins . . . generally to be found at a depth of from between 2,000 and 3,000 meters"[201] (namely, at the supposed rise, or toe, of the continental slopes) that only there exists a true geological or topographical boundary that can be clearly designated as a legal boundary. Three criticisms of this argument spring to mind. First, the 1968 Interim Report glossed over the great difference in depth and lateral extent between the 2,000- and 3,000-meter isobaths, both of which provide its empirical points of reference. Second, there cannot be much congruence with geological realities when the rise of the continental slopes is equated with the 2,500-meter bathymetric contour line since, in fact, that rise might be anywhere, on the 1968 Interim Report's own showing, between the 2,000- and 3,000-meter isobaths (a lawyer's compromise between geological facts?). Third, the Report of the United Nations Secretary General to the Economic and Social Council, *Resources of the Sea, Part One: Mineral Resources of the Sea Beyond the Continental Shelf,*[202] gives us a very different picture of the continental terraces and of the floor of the abyss. It tells us that the continental slopes extend "from the outer edge of the continental shelf to the abyssal ocean floor." Then there is the further statement: "This *abyss* or *ocean floor* appears to be a rolling plain from 3,300 to about 5,500 meters below the surface of the sea . . . . The mean depth of the superjacent waters is 3,800 metres."[203]

There is no call to judge between Dr. Pecora's work and this Report by the Secretary General of the United Nations. But when geologists and geographers do not appear to be unanimous on empirical matters, surely lawyers should hesitate to rush into the game of drawing geological boundary lines at such crushing abyssal depths as 2,500 meters, and perhaps beyond. In such regions, the ratification of a ter-

---

201. Am. Branch Comm. Interim Rep., *supra* note 173, at X.
202. U.N. Doc. E/4409/Add. 1 (mimeo. Feb. 19, 1968).
203. *Id.* at 5.

ritorial imperative in terms of a fixed isobath seems to be without practical meaning.

In the light of the foregoing points, it would be better if the outer geographical limit of an intermediate zone were the one indicated by the National Petroleum Council when it proposed that the outer edge of the geological continental slope and rise should provide "the limit of coastal-nation exclusive jurisdiction over the natural resources of the seabed and subsoil."[204] There, an empirical, topographical test was substituted, in the name of realism and practicality, for the abstract, numerical, and barely relevant test proposed by the Stratton Commission on the basis of earlier and supplanted formulations of the broad continental shelf proposals.

If the dialectic of the inclusive interests of the world community is weighed against the exclusive competences of the coastal State, the Stratton Commission's "intermediate zone" would appear to be more favorable to the coastal States than would Professor Henkin's "buffer zone" blueprint.[205] While Henkin's proposal sees the buffer zone essentially as part of the regime of the deep ocean floor, with coastal States controlling only the selection of the enterprises that may exploit the zone's mineral resources, the Stratton Commission would appear to look more to a sharing of competences and revenues. Again, while the Stratton Commission's proposal called for the registration of claims to explore and exploit the intermediate zone with its International Registry Authority "under the terms and conditions applicable to areas of the deep sea beyond the intermediate zone,"[206] the coastal State would retain effective legal control of those exploration and exploitation activities—in its capacity as the regulating State. The distinction inherent in the Stratton Commission's plan between the coastal State as the sole registering State in its intermediate zone and a State registering a claim elsewhere beyond the areas of coastal State competence lies in the coastal State's monopoly in the intermediate zone. This would provide the coastal State with a de facto source of power over activities in the zone, which would leave the international regime with less authority than under Henkin's blueprint, which calls for the exclusion of "foreign mining"—presumably unless the coastal State consents.[207]

204. NPC Report, *supra* note 96, at 67.
205. *See* STRATTON COMM'N REPORT 151–53. *See also* 3 STRATTON COMM'N PANEL REPORTS VIII–34–35.
206. STRATTON COMM'N REPORT 152.
207. L. HENKIN, *supra* note 96, at 47.

The Trusteeship Area

Chapter III of the United States Draft Convention contains an elaborate set of provisions defining coastal States' rights in the trusteeship areas beyond their continental shelves. The Draft defines the inner boundary of this area (and the outer boundary of the continental shelf) as being by straight lines "not exceeding 60 nautical miles in length" following the general direction of the 200-meter isobath.[208] Sea areas deeper than 200 meters between datum points of these lines or landward of them may be included within the continental shelf. Finally, lines longer than 60 nautical miles are permitted in cases where trenches or troughs transect these boundaries. Such lines may not be longer than "the lesser of one fourth of the length of that part of the trench or trough transecting the area 200 meters in depth or 120 nautical miles."[209] The Draft defines its proposed trusteeship area's outer boundary as "a line, beyond the base of the continental slope, or beyond the base of an island situated beyond the continental slope, where the downward inclination of the surface of the seabed declines to a gradient of 1:———."[210]

Article 26, paragraph 2, gives the trustee party the competence of determining the outer limits of its trusteeship area within the limits set out in paragraph 1 of that Draft and subject to review by its proposed International Seabed Boundary Review Commission. In the zone the coastal State is expressly prohibited from any "greater rights . . . than any other Contracting Party"[211]—"[e]xcept as specifically provided for in this chapter."[212] Despite the safeguard this formulation should provide and despite the reservation that the coastal State "acts as trustee for the international community,"[213] the coastal State is seen as enjoying, in its trusteeship area, considerably wider competences than those allowed to it under Henkin's buffer zone or the Stratton Commission's intermediate zone. The powers given to the coastal State include issuing, suspending, and revoking exploration and exploitation licenses, establishing work requirements not falling below

208. United States Draft Convention, *supra* note 190, art. 1, para. 3.
209. *Id.*
210. Art. 26, para. 1. A note attached to this article and explaining the blank space for the gradient tells us that "the precise gradient should be determined by technical experts, taking into account, among other factors, ease of determination, the need to avoid dual administration of single mineral deposits, and the avoidance of including excessively large areas in the International Trusteeship area."
211. Art. 27, para 1.
212. *Id.* The chapter referred to is Chapter III, "The International Trusteeship."
213. Art. 27, para 2.

the standards provided in the United States Draft Convention,[214] ensuring licensees' compliance with the proposed seabed Convention (and with such higher standards as the trustee State may set—these last requirements offering a further means of exclusivity in the selection of licensee enterprises), supervising its licensees and exercising civil and criminal jurisdiction with respect to exploration and exploitation activities.

One point of interest, indicating possibly hasty drafting, is the vesting of jurisdiction over the conservation of "the living resources of the seabed" and the competence to determine the "allowable catch" of these resources in the trustee State.[215] Merely by according the permitted competences to trustee States in terms of "the living resources of the seabed," Article 27, paragraph 2(h), would appear to indicate an even wider category of resources than Article 2, paragraph 4, of the Continental Shelf Convention's inclusion of "sedentary species" among the resources of the continental shelf, open-ended as that paragraph admittedly is.[216] The Draft's wording would create difficulties for anyone who wishes not to include such bottom fish as flounder, sole, and plaice among the resources of the seabed. Thus, the odd result is reached whereby coastal states may enjoy wider exclusive competences over sea bottom fisheries in their trusteeship zones than on their continental shelves.

While the coastal States administering Henkin's buffer zone are answerable for the revenues generated from the zone (subject, presumably, to a deduction of sums expended for administrative costs) to the international regime, the phraseology of the United States Draft's Chapter III and Appendix C would appear to allow an equitable apportionment of those revenues between the regime and the coastal States—the coastal States apparently being entitled to a figure between 33⅓ percent and 50 percent.[217] Thus, what started as a compromise concept allowing the coastal States little more than the privilege of deciding which enterprises should (and should not) be allowed to mine for deep seabed and subsoil resources on the periphery of their continental shelves seems, in three short steps, to have burgeoned into a regime giving coastal States beneficial rights in further seabed areas

214. The Draft's standards are specified in Appendix A.
215. Art. 27, para. 2 (h).
216. For a discussion of the open-endedness of Article 2, paragraph 4, of the Continental Shelf Convention, see Goldie, *The Oceans',* *supra* note 149, at 10–17; Goldie, *Sedentary Fisheries and Article 2(4) of the Convention on the Continental Shelf—A Plea for a Separate Regime,* 63 AM. J. INT'L L. 86, 90–91 (1969).
217. *See* United States Draft Convention, *supra* note 190, art. 28 (d) and App. C, §9 (2).

beyond those that are indisputably within their continental shelves. The third of these steps, furthermore, disguises the misuse of the term *trusteeship* in the proposal. This proposal does not provide that legal rights are accorded to the coastal State for the purpose of their fiduciary exercise for the benefit of the international community or for certain community purposes. Rather, there would appear to be a partition of both legal rights and equitable benefits between the intended beneficiaries and the trustees—an inequitable arrangement for which the Lord Chancellors would have thoroughly scraped the parties' consciences in equity cases involving trusts.

An Alternative Proposal

The buffer zones, intermediate zones, or trusteeship zones, despite matters of difference, have some fatal flaws in common. In place of a single blueprint for a regime of the buffer or intermediate zone or trusteeship area, the suggestion here is that a number of unifying forms of management—each intended to be responsive to regional community needs in specific offshore areas between the outer limits of the continental shelf properly so-called and the abyss—should be developed.[218] Different types of regimes should be formulated for different policial-geographical regions. For example, where a State possesses a coastline and offshore continental shelf formation sufficiently extensive to enable it to assert a clear claim to an appurtenant zone extending to the continental rise (so that such a claim would not be vulnerable to the possibility of being smothered in overlapping claims by adjacent or opposite States), that zone—which would, in general outline, be congruent with the Stratton Commission's intermediate zone or the United States Draft Convention's trusteeship area—could be brought within the regime of the coastal State's continental shelf. This could be done, and international claims be respected (provided agreement among nations could be achieved), by extending the affected coastal States' sovereign rights to the rise of the continental pedestal by means of a treaty. (This may appear to have a similar effect to that sought by the National Petroleum Council's suggested interpretation of the Continental Shelf Convention.) If this incorporation of the zone within the coastal State's con-

218. The argument in this section follows that in Goldie, *The United States Draft for a United Nations Convention on the International Seabed Area—A "Polite Conversation,"* 65 AM. SOC'Y INT'L L. PROC. 123, 124–27 (1971). *See also* Goldie, *Where Is the Continental Shelf's Outer Boundary?,* 1 J. MAR. L. & COMMERCE 461 (1970); Goldie, *The Continental Shelf's Outer Boundary—A Postscript,* 2 *id.* 173 (1970).

tinental shelf regime were unacceptable as benefiting only certain States, then a second possible plan for such coastal States' shore zones between the 200-meter isobath and the rise could be to place that area under the international regime for the deep ocean floor, while leaving its administration and the collection of revenue derived from exploration and/or exploitation activities under the control of the coastal State. This alternative position would leave conjoint powers with the international agency and would allocate revenue to it, or to an international fund to which the agency would also contribute from the revenues it would gain from the deep ocean itself.[219]

A third possibility might be to recognize that the zone between the 200-meter contour and the rise of the pedestal inures to the coastal State but that such a State would be specifically answerable for the administration of the zone to the international authority. This alternative would give a literal application to the term *trusteeship* by leaving legal rights with the coastal States, but assuring essential beneficial interests to the international regime. In this blueprint, the lines of accountability should be clearly laid down, so that neither the coastal State nor the international authority could erode the other's competences and rights, thereby preventing this alternative from becoming a merely temporary dispensation. A fourth compromise proposal could be designed so as to divide authority and benefits between the coastal State and the international authority. This distribution should recognize that the coastal State's sovereign rights are exercisable subject to a list of specifically defined and limited, but overriding, powers vested in the international community—for example, the power to set minimum standards for protecting the environment, or to prescribe the degree of liability for various kinds of catastrophic accidents or marine casualties, or to call for nondiscrimination against nonnational enterprises. While any of these schemes could be effectively administered off the coasts of the larger States and those with long coastlines, problems of demarcation would render them impracticable off those of small States or States with short or concave coastlines.

Where a region of small or medium-sized States exists whose individual claims (if extended into the common adjacent continental shelf borderlands) would inevitably be smothered by the overlapping claims of all the other States of the region abutting on that common shelf area, the geographical zone equivalent to the Stratton Commission's intermediate zone or the United States Draft Convention's trusteeship area could be cooperatively administered under one of the

---

219. On the Stratton Commission's International Fund proposal, see STRATTON COMM'N REPORT 147–49; 3 STRATTON COMM'N PANEL REPORTS VIII–35–38.

several models of regimes of managerial or administrative conciliation.[220] Finally, where regional political instability, territorial rivalries, irredentism, and long-term religious or racial hatreds preclude the establishment of conciliation regimes,[221] then that zone between coastal States' continental shelves and the rise of the continental slopes should be administered by the international regime established for the administration of the resources of the seabed and subsoil of the deep ocean floor.[222]

At first glance, the above group of proposals, even when the various alternatives are taken into account, would seem unduly to favor the "have" countries. They might appear to give to the States of continental or subcontinental dimensions more extensive offshore submarine regions than to middle-sized or smaller States. It should be pointed out, however, that, as a practical matter, small States or States with short or concave coastlines would not stand to get much anyway from such schemes as the Stratton Commission's intermediate zone or the United States Draft Convention's trusteeship area, and landlocked States would stand to get nothing. In fact, both classes of States would gain more from the proposal just made than from the United States trusteeship proposals. In the light of the impossibility of effectively drawing accurate seabed boundaries on the continental slopes, rise, and terraces, the latter proposal would be more likely to assure to small coastal States opposite or adjacent to one another the rights to future boundary disputes than to assure them additional resources in economically significant quantities. Secondly, not only would some "have" countries enjoy valuable increments to their offshore resources, but also a number of large and medium-sized "have-not" States—for example Nigeria, Brazil, India, Argentina, the Federation of Malaysia, and Indonesia—would qualify

---

220. For an indication of such regimes see Goldie, *supra* note 7, at 367–76.
221. *See id.* at 375–76 for a discussion of this managerial, or conciliation, regime concept in relation to submarine mineral resources.
222. The National Petroleum Council has offered the following indication of the meaning of the continental rise as the boundary between the regimes of the continental shelf and of the deep ocean floor: "Moreover, since the plunge of the slope has often been locally overlapped extensively by the sediments of the continental rise, a boundary just oceanward of the base of the slope, to include the shelf, the slope, and the landward position of the continental rise, where developed, most closely approaches the true ocean-bottom boundary between continental and oceanic areas and is the most natural and appropriate outward limit of a country's sovereign rights over bottom resources. A boundary thus drawn gives recognition to the natural oceanward extension of the domain of each coastal nation and the inclusion under its jurisdiction of that suboceanic territory over whose natural resources the coastal nation is most practically suited to exercise control." NPC Report, *supra* note 96, at 67.

to gain the addition of whatever increments to their continental shelves could be agreed upon under this blueprint.

Thirdly, these proposals are intended to bestow similar economic and organizational advantages on balkanized regions to those routinely enjoyed by federations. A managerial regime is also a supranational authority. As such, it may offer the means of removing some of the more deleterious effects of disunity in a region from the administration of offshore resources beyond the 200-meter isobath. A further advantage is that landlocked States of the region could also participate in, and benefit from, the regime, whereas they are inevitably excluded from enjoying the fruits of the powers which the trusteeship area bestows only upon coastal States.

In drafting regimes to operate as buffers between the regime of coastal States' continental shelves and a worldwide regime to govern the resources of the abyss, statesmen and lawyers could well be guided by Judge Padilla Nervo's evaluation of the continental shelf doctrine: "The purpose of the continental shelf doctrine and of the Convention is to *contribute* to a world order, in the foreseeable rush for oil and mineral resources, to avoid dangerous confrontation among States and to protect smaller nations from the pressure of force, economic or political, from greater or stronger States."[223]

These proposals are intended to assure to small States a share in the benefits that federalism can bring them with respect to areas between their continental shelves and the abyss, either directly, by encouraging federation, or indirectly, by the establishment of a supranational managerial authority over the zone between the outer limits of the continental shelf and the abyss. This writer strongly suggests that managerial regimes could more effectively "contribute to a world order" and "protect smaller nations" than could the United States Draft's recommendation of a trusteeship area.

Alternatively, another type of compromise is offered. It could consist of defining the continental shelf's outer boundary in terms of a uniform bathymetric contour line at a considerably deeper level than the Continental Shelf Convention's 200 meters, or of a measured distance from the shore as an alternative to the depth test when there is little or no continental shelf at the requisite depths. Senator Pell's proposal that the continental shelf shall be "the seabed and the subsoil of submarine areas adjacent to the coast but outside the area of the territorial sea to a depth of 550 metres or to a distance of 50 miles from the baseline from which the breadth of the territorial sea is measured, whichever results in the greatest areas of continental shelf"

---

223. North Sea Continental Shelf Cases, [1969] I.C.J. 3, 92 (Judge Padillo Nervo's emphasis).

may well provide a rational model.[224] Unfortunately, it may be rejected now as it was in 1958 at the United Nations Conference on the Law of the Sea at Geneva.[225]

Conclusion

The continental shelf doctrine is a more widely accepted basis for extending the limits of national jurisdiction than are such exotic formulations as the patrimonial sea and the other notions indicated in the introduction to this chapter. Hence, it could become a more effective instrument for converting the common heritage of mankind into the exclusive domain of encroaching States than those other formulations. For that reason, this writer urges that, while rejecting the contemporary excuses and alibis disguised behind the "creeping jurisdiction" shibboleth,[226] writers, lawyers, and diplomats give a high priority to formulating definitive criteria for delimiting the boundaries of the continental shelf in terms of such ideas as those proposed here. Failure to do so will result in ever-increasing encroachments on the free high sea both outwards and upwards, and so lead to further intensifications of the "tragedy of the [maritime] commons" as the area of free high seas becomes progressively circumscribed. Most importantly, however, the peaceful and progressive development of the emerging international law governing the distribution and uses of the ocean resources should be developed in terms of analytical concepts, not emotive slogans. While the former admit of objective meaning and investigation, the latter are not susceptible of open-minded and disinterested examination. They stand, accordingly, as the antithesis of legal thought. Nonrational and even violent elements are inherent in the appeal of their rhetoric.

224. S. Res. 33, 91st Cong., 1st Sess., art. VI (1969). The depth criterion of 550 meters that Senator Pell proposed in his S. Res. 33 was a slight amendment of his former criterion of 600 meters, which he put forward in his S. Res. 263, 90th Cong., 2d Sess. (1968), as the line of demarcation between the regime of coastal States' continental shelves and the regime of ocean space. This latter line is identical to that proposed by the United Kingdom and the Netherlands delegations in the Fourth Committee of the 1958 United Nations Conference on the Law of the Sea at Geneva. *See* Netherlands and Great Britain, Proposal, U.N. Doc. A/CONF. 13/C. 4/L. 32 (1958), U.N. Doc. A/CONF. 13/42, *supra* note 118, at 135. *See also* the Indian proposal, U.N. Doc. A/CONF. 13/C. 4/L. 29/Rev. 1 (1958), A/CONF. 13/42, *supra* note 118, at 135.

225. For the procedures and votes of the Fourth Committee on both the Indian and the Netherlands–United Kingdom proposals, see U.N. Doc. A/CONF. 13/42, *supra* note 118, at 40.

226. *See* Goldie, *supra* note 6, at 52–53.

II

Realism *vs*. Idealism as the Key to the Determination of the Limits of National Jurisdiction over the Continental Shelf

Luke W. Finlay

National Jurisdiction over the Continental Shelf

Convention on the Continental Shelf

National Jurisdiction over the continental shelf, which under Article 2 of the Convention on the Continental Shelf[1] is one of exclusive sovereign rights for the limited purpose of exploring and exploiting its natural resources, is defined in the Convention as encompassing "(a) . . . the sea-bed and subsoil of the submarine areas adjacent to the coast but outside the area of the territorial sea, to a depth of 200 metres or, beyond that limit, to where the depth of the superjacent waters admits of the exploitation of the natural resources of the said areas; (b) . . . the sea-bed and subsoil of similar submarine areas adjacent to the coasts of islands."[2]

The views of this writer[3] are in full harmony with those of the Committee on Deep Sea Mineral Resources of the American Branch of the International Law Association in that "rights under the 1958 Geneva Convention on the Continental Shelf extend to the limit of exploitability existing at any given time, within an ultimate limit of adjacency

---

1. Convention on the Continental Shelf, Apr. 29, 1958, art. 2, [1964] 1 U.S.T. 471, T.I.A.S. No. 5578, 499 U.N.T.S. 311.

2. Art. 1.

3. *See* Finlay, *The Outer Limit of the Continental Shelf—A Rejoinder to Professor Louis Henkin,* 64 AM. J. INT'L L. 42 (1970); Finlay, *The National Interest and the Limits of the Continental Shelf,* 4 MARINE TECHNOLOGY SOC'Y J. 71 (1970); Finlay, *Rights of Coastal Nations to the Continental Margins,* 4 NATURAL RESOURCES LAWYER 668 (1971). See also Am. Bar Ass'n. Resolution re Natural Resources of the Sea, adopted by the House of Delegates Aug. 6, 1973, and the Report of the Section of Natural Resources Law Recommending Adoption of that Resolution, for which this writer acted as rapporteur, 6 NATURAL RESOURCES LAWYER 589 (1973).

which would encompass the entire continental margin."[4] The view that the Convention should be broadly construed has gained wide support among objective scholars, both in the United States and abroad.[5] Recently, however, critics have characterized this interpretation as overreaching and have asserted that if it were adopted and acted upon by the United States, it would be regarded as a "grab," even if all coastal nations followed suit.[6] A few years ago the question was not whether the Convention should be given a broad or a narrow interpretation but whether there was any limit on exploitability short of mid-ocean.[7] This mid-ocean or "national lake" theory is no longer seriously discussed by objective scholars, and no serious effort was made either to interpret or to amend the Convention to prescribe narrow limits of national jurisdiction until the potential

4. Comm. on Deep Sea Mineral Resources of the Am. Branch of the Int'l L. Ass'n, Second Interim Report, July 1970, at 1. As pointed out in the Committee's earlier report, Comm. on Deep Sea Mineral Resources of the Am. Branch of the Int'l L. Ass'n, Interim Report, July 19, 1968, at XI, XVIII, in special cases, such as those of very narrow or ill-defined continental margins and semienclosed seas, exploitability, as qualified by adjacency, may extend coastal State jurisdiction beyond the continental margin. References in this paper to the primary aspect of the definition should not be understood to negate this latter aspect, with which the author also concurs.

5. Jennings, *Jurisdictional Adventures at Sea—Who Has Jurisdiction over the Natural Resources of the Seabed?* 4 NATURAL RESOURCES LAWYER 829, 832–33 (1971); F. GARCIA AMADOR, THE EXPLOITATION AND CONSERVATION OF THE RESOURCES OF THE SEA 130 (2d ed. 1959). *See also* M. McDOUGAL & W. BURKE, THE PUBLIC ORDER OF THE OCEANS 683 (1962); Am. Bar Ass'n. Resolution and the Report recommending its adoption, note 3 *supra*. Professor Jennings has expressed the interesting view that jurisdiction has become vested in nonsignatories of the Convention but may be dependent upon exploitability as regards the relations of the signatories *inter'sese*, unless this requirement may now be regarded as *functus officio*, having been overtaken by customary law. Jennings, *The Limits of Continental Shelf Jurisdiction: Some Possible Implications of the North Sea Case Judgment*, 18 INT'L & COMP. L.Q. 819, 831–32 (1969). *See also* Nat'l Petroleum Council, Interim Report on Petroleum Resources Under the Ocean Floor, July 9, 1968, at 6; Nat'l Petroleum Council, Report on Petroleum Resources Under the Ocean Floor, March 1969, at 10, 13, 55 *et seq.;* Nat'l Petroleum Council, Supplemental Report on Petroleum Resources Under the Ocean Floor, March 1971, at 6. For the views of the American Petroleum Institute, see Finlay, *The Draft United Nations Convention on the International Seabed Area—American Petroleum Institute Position*, 4 NATURAL RESOURCES LAWYER 73, 75 (1971).

6. *See, e.g.,* COMM'N ON MARINE SCIENCE, ENG'R & RESOURCES, OUR NATION AND THE SEA 145 (1969) [hereinafter cited as the STRATTON COMM'N REPORT].

7. *See, e.g.,* S. ODA, INTERNATIONAL CONTROL OF SEA RESOURCES 167 (1963); C. FRANKLIN, THE LAW OF THE SEA: SOME RECENT DEVELOPMENTS 25, 29 (1961). *See also* Creamer, *Title to the Deep Seabed: Prospects for the Future*, 9 HARV. INT'L L.J. 205, 212–13 (1968).

wealth of the seabed began to arouse interest as a possible source of revenue for international community purposes.[8]

International Law Commission View

The International Law Commission (ILC), charged by the United Nations General Assembly with the task of codifying the law of the sea, changed its view on the extent of national jurisdiction over the continental shelf several times. In 1951 the ILC favored the test of exploitability alone; in 1953 it shifted to the 200-meter water-depth limit alone; and in 1956 its final recommendation combined the two prior positions in basically the same language as the 1958 Convention.[9] Throughout these prolonged deliberations, there was never any suggestion that a broad definition would trespass on "the common heritage of mankind."[10] On the contrary, the question was the very simple one of whether it would be preferable to have a definition broad enough to anticipate the future needs of the coastal States in light of possible technical advances or to deal only with present needs and reexamine the seaward extent of coastal jurisdiction only later if technical advances should require the fixing of a greater depth.[11] No other conclusion may be drawn from the following illuminating passage in the 1956 *Yearbook* of the ILC:

Mr. FRANÇOIS, Special Rapporteur, said that the Commission [in 1953] had fixed the limit of 200 metres merely in order to prevent each State from claiming a continental shelf of whatever size it wished. . . .

8. Jennings, *The Limits, supra* note 5, at 821. It is more than a bit surprising that advocates of a narrow shelf continue to drag this red herring across the trail unless their purpose is to use it as a scare tactic rather than as a serious legal discussion. *See, e.g.,* W. FRIEDMANN, THE FUTURE OF THE OCEANS 4–6 (1971). As Jennings observed in the cited article, this extreme view of the effect of exploitability, which completely ignores the companion criterion of adjacency, was effectively laid to rest when the United Nations General Assembly, on December 17, 1970, adopted by a vote of 108 to 0, with 14 abstentions, a Declaration of Principles "[a] *ffirming* that there is an area of the sea-bed and the ocean floor, and the subsoil thereof, beyond the limits of national jurisdiction, the precise limits of which are yet to be determined. . . ." G.A. RES. 2749, 25 U.N. GAOR Supp. 28, at 24, U.N. Doc. A/8097 (1970).

9. For a detailed discussion of these developments, see 4 M. WHITEMAN, DIGEST OF INTERNATIONAL LAW 829–38 (1965); M. McDOUGAL & W. BURKE, *supra* note 5, at 672–80.

10. This phrase was borrowed from Declaration of Principles, G.A. RES. 2749, note 8 *supra.*

11. [1953] 2 Y.B. INT'L L. COMM'N 46, U.N. Doc. A/CN.4/SER. A/Add. 1 (1953).

As for the continental terrace, exclusive rights of exploitation of that part of it which lay at a depth of less than 200 metres were already recognized under the Commission's draft articles. The question of the right to exploit any parts of it which lay at a greater depth was of no significance, since such exploitation was for the moment physically impossible. *The Commission had, however, admitted that if any State could demonstrate the possibility of exploiting the sea-bed at a greater depth, the limit of 200 metres could not be retained.*[12]

The Development of United States Views

The ILC attitude as expressed by François applied equally to the 1945 Truman Proclamation on the Continental Shelf.[13] The Proclamation asserted United States jurisdiction over the natural resources of the seabed and subsoil of the continental shelf without any further definition of that term. The White House press release announcing the issuance of the proclamation, however, stated *inter alia* that the United States generally viewed submerged lands contiguous to the continent and covered by no more than 100 fathoms (600 feet) of water as the continental shelf. The Legal Adviser of the Department of State, Herman Phleger, referring to the 100-fathom figure in an address to the American Branch of the International Law Association on May 13, 1955, virtually parroted the words of François: "This limitation—defined in article 1 of the International Law Commission's [1953] draft as 200 meters—would seem to cover all practical needs for the foreseeable future and to have the advantage of definiteness. If future technical advances should render this formulation inadequate, it can be reconsidered in the light of intervening experience."[14]

The Committee on Territorial Waters and United Nations Activities of The Maritime Law Association of the United States undoubtedly had these views in mind when, in 1957, it recommended general support of the final (1956) ILC Draft Articles on the continental shelf with minor revisions. Thus, after quoting the proposed dual definition of the continental shelf in terms of both depth and exploitability, the Committee stated:

12. [1956] 1 Y.B. Int'l L. Comm'n 138, U.N. Doc. A/CN.4/SER. A (1956) (emphasis added).
13. Presidential Proclamation No. 2667, Policy of the United States With Respect to the Natural Resources of the Subsoil and Sea Bed of the Continental Shelf, Sept. 28, 1945, 3 C.F.R. 67 (1943–1948 Comp.). This proclamation and pertinent extracts from the press release are reproduced in 4 M. Whiteman, *supra* note 9, at 756–58.
14. Address by Herman Phleger, Legal Adviser, Dept. of State, "Recent Developments Affecting the Regime of the High Seas," Am. Branch of the Int'l L. Ass'n, New York, N.Y., May 13, 1955, 4 M. Whiteman, *supra* note 9, at 762.

The Truman Proclamation of 1945 spoke without amplification of "continental shelf," apparently using the term in its geological sense as the continuation of the land mass out to the point where the continental slope, falling off steeply to great depths, begins. The definition adopted by the International Law Commission seems desirable for at least two reasons: (1) The continental shelf has developed a broader connotation for legal and political purposes than its technical connotation among geologists; and (2) although the 200 metre delimitation is clearly adequate for present-day exploitation potential, it seems wise, *since the crux of the doctrine from the start has been exploitability,* to allow for future technological advances which may expand exploitation horizons.[15]

The 1956 decision of the ILC to incorporate the dual test of the 200-meter water depth and exploitability in its final definition of the continental shelf stemmed from a decision taken by the United States and nineteen other American States at the Ciudad Trujillo Conference only a few weeks previously. At the Conference it had been unanimously resolved that

[t]he sea-bed and subsoil of the continental shelf, continental and insular terrace [which the conference report defined as the continental shelf and the continental slope "from the edge of the shelf to the greatest depths"], or other submarine areas, adjacent to the coastal state, outside the area of the territorial sea, and to a depth of 200 meters or, beyond that limit, to where the depth of the superjacent waters admits of the exploitation of the natural resources of the sea-bed and subsoil, appertain exclusively to that state and are subject to its jurisdiction and control.[16]

A majority of the members of the ILC were opposed to the inclusion of a reference to the continental terrace in their definition, but they did vote in favor of the generic term *submarine areas adjacent to the coast,* with the dual tests of the 200-meter water depth and exploitability beyond that depth as the basis for determining the outer boundary of the areas covered.[17] Myres McDougal and William Burke refer to the Ciudad Trujillo Conference as having had a "decisive effect upon the discussions in the International Law Commission and its ultimate recommendation."[18] They further state that,

15. Maritime Law Ass'n, Report of Comm. on Territorial Waters and U.N. Activities, Doc. No. 408, at 4159, Oct. 24, 1957 (emphasis added).
16. Inter-American Specialized Conference on "Conservation of Natural Resources: The Continental Shelf and Marine Waters," *Final Act,* Ciudad Trujillo, March 15–28, 1956. The definition of *continental terrace* is at 34.
17. 1 Y.B. INT'L L. COMM'N, *supra* note 12, at 130–31, 139–40.
18. M. MCDOUGAL & W. BURKE, *supra* note 5, at 680. *See also* Whiteman, *Conference on the Law of the Sea: Convention on the Continental Shelf,* 52 AM. J. INT'L L. 629, 633 n.21 (1958).

without a specific reference to the continental terrace, the language adopted by the ILC had the same effect as that of the Ciudad Trujillo Resolution.[19] The Chairman of the 1956 Session of the ILC, F. V. Garcia Amador, expressed identical views.[20]

Before the adoption of the final text of the 1958 Convention on the Continental Shelf, a vote had been taken in Plenary Session on whether to retain the dual test of the 200-meter water depth and exploitability or to strike out exploitability and use the 200-meter water depth alone as the outer limit of national jurisdiction over the continental shelf. The United States and Soviet Union Delegations were among those favoring the dual test, which was retained in the Convention by a vote of 48 to 20 with 2 abstentions.[21] Thus, Ambassador Arthur H. Dean, head of the American Delegation at Geneva, said before the Senate Committee on Foreign Relations at the January 1960 hearings on advice and consent to ratification of the Convention: "The clause which protects the right to utilize advances in technology at greater depths beneath the oceans was supported by the United States and was in keeping with the inter-American conclusions at Ciudad Trujillo in 1956. It was included in the ILC 1956 draft."[22]

In the years following ratification of the Convention, all departments of the Executive Branch of the United States uniformly accepted a broad definition of the Convention but made no effort—nor was there a practical necessity therefor—to pinpoint the outer limits of national jurisdiction. As may be seen from Figure 1, the original edition of Department of State Geographic Bulletin Number 3 clearly stands for a concept of exploitability extending national jurisdiction "to indefinite distance" beyond the outer edge of the physical continental shelf.[23]

While this indefinite distance view prevailed in the Department of State, the Department of the Interior was issuing leases off the coast of California in water depths ranging to 4,000 feet and publishing lease maps for areas off the same coast with water depths as great as 6,000 feet. Frank J. Barry, Solicitor for the Department of the Interior, reported on these developments at length during the American Bar

19. M. McDougal & W. Burke, *supra* note 5, at 683.
20. F. Garcia Amador, note 5 *supra*.
21. 2 U.N. Conf. on the Law of the Sea, Geneva 1958, Official Records, (*Plenary Meetings*)13, U.N. Doc. A/CONF. 13/38 (1958).
22. *Hearing on Conventions on the Law of the Sea Before the Senate Comm. on Foreign Relations,* 86th Cong., 2d Sess. 108 (1960); 42 Dep't State Bull. 258 (1960).
23. U.S. Dep't of State, Pub. No. 7849, at 7–8, 28 (1965).

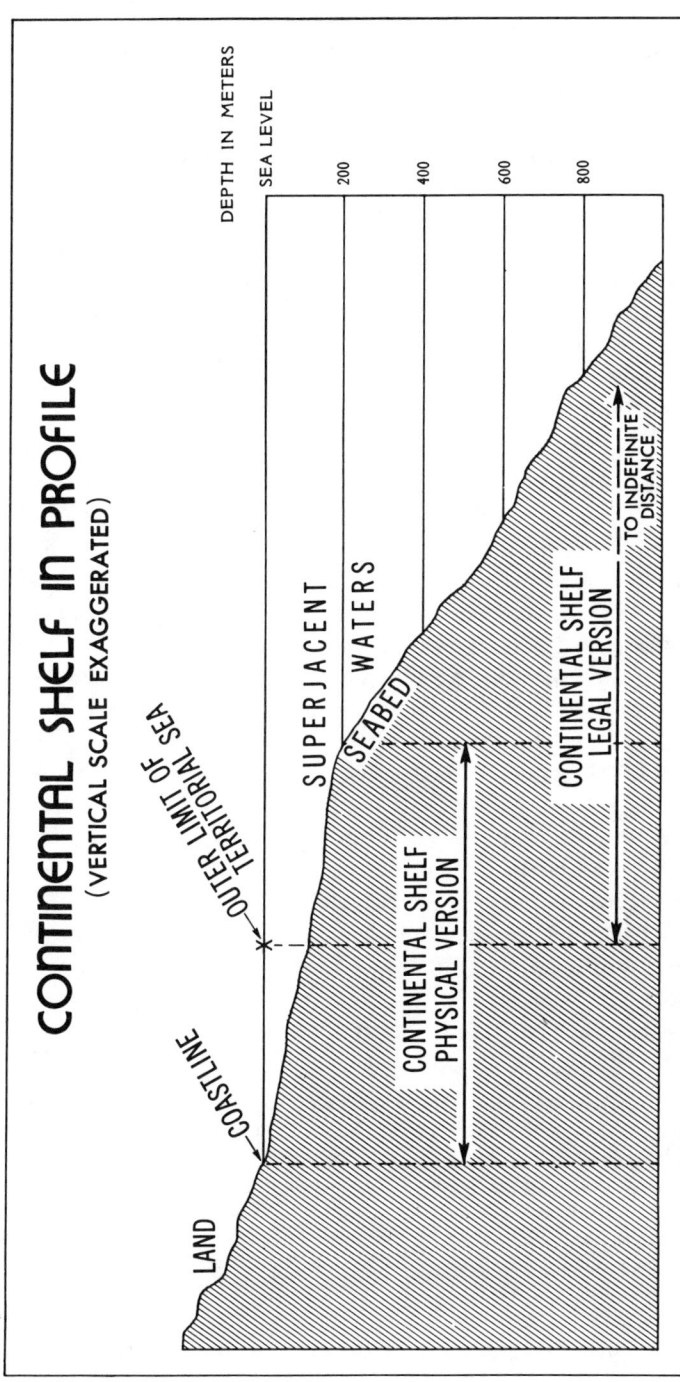

Source: *Sovereignty of the Sea*, Geographic Bulletin No. 3 (April 1965), Department of State Publication 7849.

Fig. 1. Continental shelf in profile

Association National Institute on Marine Resources in June 1967.[24]
He made particular mention of his office's 1961 opinion approving a
lease of land on the outer continental shelf for the dredging of phos-
phorite nodules lying on the ocean floor forty miles off the coast of
California, in an area known as Forty-Mile Bank. He said:

The water depth in that area is from 240 to 4,000 feet. Most was at a depth of
far greater than 600 feet. It is separated from the mainland by an ocean floor
trench as much as 4,000 to 5,000 feet deep. The Opinion concluded that the
Act permitted leasing in that area because the ratification of the Convention
by the United States constituted an assertion of rights to the seabed and
subsoil as far seaward as exploitation is possible. The Opinion was submitted
to the Departments of State and Justice to determine whether they had objec-
tions and they registered none.[25]

He also mentioned that in February 1967 his office had advised the
United States Army Corps of Engineers that the Department of the
Interior clearly regarded the Cortez Bank as being within the jurisdiction of the United States under the Outer Continental Shelf Lands
Act[26] and the Geneva Convention. This bank is located about fifty
miles from San Clemente Island and a hundred miles from the mainland, being seaward of, and on the same prolongation of the California
landmass as, the Forty-Mile Bank. The water over the Cortez Bank is
only 2½ feet deep at its shallowest point, but it is separated from the
mainland and San Clemente Island by ocean floor trenches as much
as 4,000 to 5,000 feet deep. Barry observed: "The area was covered by
leasing maps regarded as an affirmative assertion of jurisdiction by
the United States and by the emplacement of a Coast Guard buoy.
Additionally, a published scientific report showed the Bank to be an
extension of the land mass of Southern California."[27]

He then explained the Department's leasing policy: "You may want
to know whether the Department has decided on a line beyond which it
will not lease, or whether it has decided to lease as far out as anyone
might suggest. The answer on both counts is no. Each case will be

---

24. Barry, *The Administration of the Outer Continental Shelf Lands Act*, 1 NATURAL
    RESOURCES LAWYER 38–48 (1968). The Outer Continental Shelf Lands Act, it may
    be noted, applies to all submerged lands lying seaward of those ceded to the United
    States coastal states by the Submerged Lands Act, the seabed and subsoil of which
    "appertain to the United States and are subject to its jurisdiction and control. . . ."
    43 U.S.C. §1331(a) (1970). As Mr. Barry indicated in his paper, this has been con-
    strued in harmony with the Convention, to which the United States is a party.
25. Barry, *supra* note 24, at 46.
26. 43 U.S.C. §§1331 *et seq.* (1970).
27. Barry, *supra* note 24, at 47.

considered individually, with consultation with the State and Justice Departments where appropriate."[28]

A statement by the late William T. Pecora, then Director of the United States Geological Survey, to the Senate Subcommittee on Air and Water Pollution on February 28, 1969, indicates, however, that the position of the department had sharpened somewhat since Barry's remarks. Secretary of the Interior Hickel, having been asked the total area of the outer continental shelf, passed the question to Pecora, who replied: "The Outer Continental Shelf, and we define it, *meaning the shelf and slope,* has about 1.2 million square miles. Now within that area, within the area of the Outer Continental Shelf under Federal control, less than 1 percent is out under lease at the present time."[29]

On March 27, 1969, Pecora made the following enlightening remarks to the Subcommittee on Minerals, Materials and Fuels of the Senate Committee on Interior and Insular Affairs on the delimitation of the continental shelf:

> For many years the term Continental Shelf has been used in a general sense to imply those terraced submerged lands off the continent that go to about a depth of 200 meters. In a geological sense this is only one part of the topographic surface. The continental slope which is seaward from the shelf is part of the continental terrace and is composed of the same rocks.
>
> *Therefore, in any concept of the outer limit of the continental jurisdiction from a scientific and geological sense we would recommend to the Senate that there is a continuity seaward to the base of the continental slope as a minimum.*
>
> This is a rather important point because one section of the Marine Commission has recommended that for international relationships . . . we consider merely the 200 meter line or 50 miles, whichever is the greater.
>
> In fact the base of the continental slope reaches a depth in excess of 2,500 meters and sometimes in some places goes well beyond the 50-mile line.[30]

The United States Coast and Geodetic Survey's now retired au-

28. *Id.*
29. *Hearings on S. 7 and S. 542 Before the Subcomm. on Air & Water Pollution of the Senate Comm. on Public Works,* 91st Cong., 1st Sess., pt. 4, at 942, 961 (1969) (emphasis added).
30. STAFF OF SENATE COMM. ON INTERIOR & INSULAR AFFAIRS, 91ST CONG., 1ST SESS., SELECTED MATERIALS ON THE OUTER CONTINENTAL SHELF 42 (Comm. Print 1969) (Memorandum by the Chairman to Members of the Comm. on Interior and Insular Affairs) (emphasis added). Menard and Smith treat the continental shelf and the continental slope as a single province, explaining that "[s]helf and slope are grouped because they are merely the top and front of the margins of continental blocks." Menard & Smith, *Hypsometry of Ocean Basin Provinces,* 71 J. GEOPHYSICAL RESEARCH 4308 (1966).

thority on sea boundaries, Aaron L. Shalowitz, has consistently construed the Convention as confirming broad coastal State jurisdiction
over the continental margin.[31] Similarly, that the Department of the
Navy for some years accepted a broad interpretation of coastal State
jurisdiction over seabed resources is understandable in view of the
leasing practices of the Secretary of the Interior as administrator of
the provisions of the Outer Continental Shelf Lands Act and the fact
that the United States Navy's own representatives at Ciudad Trujillo
had encouraged United States approval of an affirmative vote on the
exploitability aspect of the resolution adopted at the Conference.[32] In
August 1967 the Judge Advocate General of the Navy sent a team of
officers to the annual meeting of the American Bar Association to
make a five-part presentation that clearly demonstrated the Navy's
acceptance of the possibility of, and its ability to live with, a broad interpretation of coastal State jurisdiction under the Convention.[33] The
same view was expressed by officials of the Department of the Navy as
recently as 1968.[34]

After the views of the Stratton Commission had started to
crystallize in favor of a narrow redefinition of the continental shelf,
the Navy's views began to change; and, in the summer of 1969, G.
Warren Nutter, Assistant Secretary of Defense for International Security Affairs, advised the Subcommittee on Ocean Space of the
Senate Committee on Foreign Relations that the Department of
Defense favored a narrow boundary. Nutter, however, declined to
suggest a precise mileage or depth pending resolution of the interde-
partmental discussions then in progress.[35] These discussions, of
course, eventually led to the President's policy statement of May 23,
1970, under which leases of continental shelf areas beyond the 200-
meter water depth would be issued subject to the international regime

31. *See* A. SHALOWITZ, SHORE AND SEA BOUNDARIES 255, n.111 (1962); Letter from A.
    Shalowitz to the Editor, N.Y. Times, May 31, 1970, §3, at 15, col. 5; Letter from A.
    Shalowitz to the Editor, Wash. Post, Sept. 12, 1971, §B, at 7, col. 2.
32. 3 STRATTON COMM'N PANEL REPORTS, MARINE RESOURCES AND LEGAL-POLITICAL
    ARRANGEMENTS FOR THEIR DEVELOPMENT VIII–20, n.69 (1969).
33. *The Fourth Dimension of Seapower—Ocean Technology and International Law,* 22
    JAG J. 26, 38, 39–41 (1967).
34. Frosch, *Marine Mineral Resources—National Security and National Jurisdiction,*
    in Proceedings of a Symposium on Mineral Resources of the World Ocean 1968 at
    96, 99–101 (Grad. Sch. of Oceanography, U. of R. I., Occasional Publication
    No. 4).
35. *Hearings on S. 33, Activities of Nations in Ocean Space, Before the Subcomm. on
    Ocean Space of the Senate Comm. on Foreign Relations,* 91st Cong., 1st Sess. 235
    (1969).

to be agreed upon.[36] Moreover, it is apparent from an examination of the October 1969 revision of the Department of State Geographic Bulletin Number 3 that the Department of State had also modified its earlier views.[37]

Regardless of what was behind the shift in the views of the Departments of State and Defense, it seems apparent that their earlier views and those of Interior and Commerce are far more meaningful in arriving at a sound interpretation of national rights under the Convention. It is equally apparent to this writer that the potential of the continental slope as a source of funds for international community purposes is of no probative value whatever in interpreting the rights of coastal States under existing international law. Nor is it likely, for reasons soon to be explained, to have meaningful influence in the achievement of a viable international treaty redefining the limits of national jurisdiction over the continental shelf. As Professor R. Y. Jennings very aptly commented: "It is said that politics is the art of the possible. The same may be said of international law. So the twin questions we have to look at are, what is the existing law; and what kind of new regime and what new limits are both desirable and obtainable?"[38]

The Arguments over Redefinition of the Continental Shelf

The principal arguments advanced in support of a narrow redefinition of the continental shelf fly squarely in the face of Jennings's sound common-sense suggestion, and the balance of this chapter will be devoted to their refutation.

International Seabed Regime

It is generally agreed that existing international law governing the exploration and exploitation of the resources of the seabed and subsoil of the high seas beyond the limits of national jurisdiction provides inade-

36. 9 INT'L LEGAL MATERIALS 806 (1970).
37. U.S. Dept. of State, Pub. No. 7849 (1965, *as revised* Oct. 1969, *and released* Dec. 1969). The chart reproduced in the revised publication is unchanged, but the text was changed to reflect doubt on the meaning of Article 1 of the Convention, including the possibility that exploitability is limited to the physical continental shelf. *Id.* at 8–9. This interpretation was expressly negated by the ILC in its 1956 Report to the U.N. General Assembly, Int'l L. Comm'n Report, [1956] 2 Y.B. INT'L L. COMM'N 296–297, 11 U.N. GAOR Supp. 9, at 41–42, U.N. Doc. A/3159 (1956).
38. Jennings, *Jurisdictional Adventures at Sea, supra* note 5, at 829.

quate security of tenure to warrant the huge investments that would be involved in such activities.[39] It does not follow, however, that the filling of this need provides the opportunity for a de novo examination of the limits of national jurisdiction over the continental shelf, despite the urgent plea of authors such as Wolfgang Friedmann for a scheme of international redistribution of seabed resources.[40]

There are entirely too many vested rights in the resources of the seabed and ocean floor of the continental margin beyond the 200-meter isobath to make an examination de novo of those limits remotely feasible. A recent survey indicates that, including colonies and protectorates, 111 free-world political entities have awarded offshore concessions or leases and that 55 of these have done so in waters extending at least in part beyond the 200-meter isobath and, in a few instances such as Canada and South-West Africa, into depths of 3,000 meters or more.[41] Since the time when the first true offshore platform was built in fifty feet of water off the Louisiana coast in 1947, some 20,000 wells were drilled off the coasts of seventy countries and accounted for one-sixth of the world's oil production as of 1972, a percentage that is still increasing.[42] Professor Jennings believes that, as a consequence of this widespread State practice, his a priori conclusion that the coastal States have exclusive jurisdiction over the continental slope and the subsoil of that part of the natural prolongation of their land mass is in the process of confirmation as a matter of customary international law.[43]

Moreover, there has been a recent sharp increase in the number of States seeking the protection of the Convention. As of August 18, 1967, when the question of the reexamination of the law of the sea was first brought before the United Nations by the introduction of the

---

39. *Id.* at 833. As the STRATTON COMM'N REPORT, *supra* note 6, at 146, observes, a nation may explore and keep what it finds under present law, but there is no agreement among international lawyers that it may also exclude poachers.

40. *But see* Friedmann, *supra* note 8, at 16.

41. These figures are from an unpublished survey made in the summer of 1972. In a paper on *Offshore Petroleum Development and Potential* presented to the Columbia University Seminar on Uses of the Ocean on February 7, 1974, Dr. H. R. Gould stated (at page 9) that according to the Foreign Scouting Service of Petroconsultants, S.A., some 50 nations have now opened acreage beyond their continental shelves to petroleum operations, in some instances to water depths as great as 12,000 feet.

42. The New Adventurers [Report of the Standard Oil Company of New Jersey (1972)] [now Exxon Corporation]. Dr. Gould reported that as of February 1974 offshore oil accounted for about 18 percent of total world and U.S. production, Gould, *supra* note 41, at 3.

43. Jennings, *The Limits of Continental Shelf Jurisdiction, supra* note 5, at 830.

Malta Resolution,[44] there were only thirty-seven adherents to the Convention. Between that date and the introduction of the United States Draft of a proposed United Nations Convention on the International Seabed Area on August 3, 1970,[45] four additional States acceded to the Convention: Thailand, Trinidad and Tobago, Kenya, and Canada. With the introduction of the Draft Treaty, the pace quickened and another thirteen States—Mauritius, Spain, Tonga, the Republic of China (Taiwan), Fiji, Norway, Swaziland, Nigeria, Costa Rica, Greece, Lesotho, the German Democratic Republic, and Cyprus[46]—have become parties, bringing the total to 54. These sixteen newcomers to the Convention include several with extensive claims to continental margins, particularly Canada, Nigeria, and Norway, the logical conclusion being that they wished to fortify their rights under customary international law by bringing themselves within the specific protection of the exploitability clause of the Convention.

The consequences of a new international seabed treaty as regards nonparty States will be discussed below,[47] but it is abundantly clear that in the deliberations now in progress under United Nations auspices, the boundary of the legal continental shelf must be approached in the light of the existing realities and not, as the World Federalists would prefer it, as if ocean space were a tabula rasa.

Uncertainty of Continental Shelf Boundaries

Some argue that the Convention's definition of the boundary of national jurisdiction over the continental shelf is so uncertain that it constitutes no agreement at all. Representative of this approach is that of former Defense Department official Leigh S. Ratiner, who characterizes the Convention as a failure and its definition of the limits of national jurisdiction as a "non-boundary" decision.[48] The obvious implication of such remarks is that the Convention is so vague as to be unenforceable and that the parties to the Convention are forced to negotiate for a new definition.

The definition of Article 1 of the Convention is admittedly couched in general terms, but intentionally so. In the first place, it must be

---

44. U.N. Doc. A/6695 (1967).
45. U.N. Doc. A/AC.138/25 (1970), 9 INT'L LEGAL MATERIALS 1046–80 (1970).
46. For the latest list of adherents, see Treaties in Force for the United States as of Jan. 1, 1974, U.S. DEP'T OF STATE, Pub. No. 8755 (1974); Accession of Cyprus is reported in 13 INT'L LEGAL MATERIALS 1046 (1974).
47. *See* pp. 92–96.
48. Ratiner, *United States Ocean Policy—An Analysis,* 2 J. MARITIME L. & COMMERCE 225, 230–31 (1971).

borne in mind that the legal continental shelf is a province of the high
seas, and as such, unlike the physical continental shelf, which begins at
the coastline, it must have as its starting point the outer limit of the
territorial sea. This means that throughout the world a part of the
physical continental shelf is landward of the legal continental shelf and
that, in localities such as portions of the coast off Chile and Alaska,
where the coastline is extremely precipitous, the entirety of the
physical continental shelf is landward of the legal continental shelf.
Moreover, the 200-meter isobath itself is not tied to any topographic
feature. While it is loosely—and inaccurately—said that 200 meters is
the average depth of the outer edge of the physical continental shelf, in
reality the average depth is only 130 meters and the range in various
parts of the world is from 50 to 550 meters.[49] Thus, the immediately
vested rights of the coastal States under the Convention to a depth of
200 meters, regardless of exploitability, extend for at least some
distance down the continental slope in most parts of the world. Finally, there was the firm demand of the Latin Americans, acceded to
by the ILC in 1956 and ratified by the Geneva Conference in 1958,
that, in point of law at least, the States with precipitous coastlines
should be accorded equitable treatment by being assured of exclusive
seabed-resource jurisdiction even beyond the continental margin
within the limits of adjacency and exploitability.[50] It was these factors,
coupled with the unwillingness of a majority of the members of the
ILC to consider a formula based on a greater depth than 200 meters
or a specific distance from shore,[51] that led to agreement on the
phrase "submarine areas adjacent to the coast but outside the area of
the territorial sea" as the definition of the areas to be covered, with
immediate vesting to a depth of 200 meters and future vesting beyond
that depth as exploitability warranted it.

The particular function of international lawyers and jurists is to im-
plement, not to negate, the intention of treaty makers. The issue has
never arisen as to the precise scope of the equitable treatment designed for States with precipitous coastlines. As previously noted,
however, the preponderance of authority supports the view that, in its
major thrust, the Convention confirms the exclusive jurisdiction of the
coastal States over the seabed resources of the entire continental
margin as advances in technology make possible their exploitation.[52]

49. NPC Report, *supra* note 5, at 28 (Fig. 3, A), 105.
50. F. GARCIA AMADOR, note 5 *supra*.
51. *See* Whiteman, *supra* note 18, at 634, for a discussion of rejected criteria.
52. *See* notes 5 and 31 *supra*. For an analysis of Henkin's differing views, see Finlay *The
    Outer Limit of the Continental Shelf*, note 3 *supra*. *But cf.* Henkin, *A Reply to Mr.
    Finlay*, 64 AM. J. INT'L L. 62 (1970).

No one at the 1958 Geneva Conference believed that the Fourth Committee had failed to accomplish its mission of coming up with an effective treaty on the continental shelf. The following comment by the delegate from Colombia, after final agreement had been reached on the language of Article 1, evoked no contrary view and seemingly expressed the sense of the Committee:

The article provided a permanent solution for the problem of the definition of the continental shelf and for that of scientifically possible exploitation; safeguarded the States without a wide shelf, while in no way causing prejudice to those States that had such a shelf, because the latter would have unimpaired enjoyment of their shelf; guaranteed equality of rights to all coastal States, which would enjoy equal rights in the submarine area, since they would stem from the basic right of self-preservation and defense of States.[53]

Had the contrary been the case, one would have expected the outer limit of the continental shelf to have been included with the unresolved issues of the breadth of the territorial sea and coastal State fisheries rights for further consideration at the follow-up 1960 Geneva Conference on the Law of the Sea. Yet the record fails to reveal the slightest indication of such a need from any source.[54]

Finally, it is impossible to conceive of the International Court of Justice's having given its wholehearted approbation to Article 2 of the Convention in the following ringing language had it had the slightest reservation regarding the legal adequacy of that Article and the companion definition of Article 1:

[W]hat the Court entertains no doubt is the most fundamental of all the rules of law relating to the continental shelf, enshrined in Article 2 of the 1958 Geneva Convention, though quite independent of it,—namely the rights of the coastal State in respect of the area of continental shelf that constitutes a natural prolongation of its land territory into and under the sea exist *ipso facto* and *ab initio,* by virtue of its sovereignty over the land, and as an extension of it in an exercise of sovereign rights for the purpose of exploring the seabed and exploiting its natural resources. In short, there is here an inherent right. In order to exercise it, no special legal process has to be gone through, nor have any special legal acts to be performed. Its existence can be declared (and many States have done this) but does not need to be constituted. Furthermore, the right does not depend on its being exercised. To echo the

53. 6 U.N. CONF. ON THE LAW OF THE SEA, GENEVA 1958, OFFICIAL RECORDS (FOURTH COMM.) 41, U.N. Doc., A/CONF. 13/42 (1958).
54. *See* the resolution adopted at the 1958 Conference on the desirability of convening a second conference for further consideration of the questions left unsettled in 1958. U.N. Doc. A/CONF. 13/L. 56 (1958). *See also* the remarks of the representatives of Cuba, Ceylon, and Australia in Plenary Session at the close of the 1958 Conference, *supra* note 21, at 73–75.

language of the Geneva Convention, it is "exclusive" in the sense that if the coastal State does not choose to explore or exploit the areas of shelf appertaining to it, that is its own affair, but no one else may do so without its express consent.[55]

Some advocates of a narrow redefinition of the continental shelf rely on a statement elsewhere in the opinion of the Court that "by no stretch of imagination can a point on the continental shelf situated say a hundred miles or even much less, from a given coast, be regarded as 'adjacent' to it, or to any coast at all, in the normal sense of adjacency, even if the point concerned is nearer to some one coast than to any other."[56] In so doing, they are guilty of an error that any first-year law student should be able to detect. A study of the opinion, and of the charts incorporated in it, shows that the Court was here speaking, not of the seaward extent of coastal State jurisdiction, but of adjacency as a test for fixing the lateral boundary between contiguous States, a concept that the Court rejected. The natural prolongation of the land mass was the Court's guideline for determining the seaward extent of coastal State jurisdiction and it laid down the principles to be followed by the three litigant States before it—the Netherlands, the Federal Republic of Germany, and Denmark—in dividing among themselves seabed resources of the North Sea extending in excess of 150 nautical miles from shore to the common boundary with the equally broad continental shelf of the United Kingdom.[57] A suggestion that the important discoveries of oil and gas in portions of the North Sea far more than 100 miles from the nearest coast, such as the Ekofish Field in the southwest corner of the Norwegian sector,[58] are the rightful property of the international community would be met with derision in the capitals of the countries bordering on the North Sea.

55. North Sea Continental Shelf Cases, [1969] I.C.J. 3, 22, 8 INT'L LEGAL MATERIALS 340, 357 (1969). This statement of the Court, at paragraph 19, fully bore out the view of Marjorie Whiteman, representative of the United States to Committee IV (Continental Shelf) at the 1958 Geneva Conference, that the Convention would have an important influence on the law of the continental shelf whether or not it attracted a great number of States as parties. Whiteman, *supra* note 18, at 659.
56. [1969] I.C.J. at 30.
57. *Id.* at 15 (Map 3). The distance to the common boundary with the continental shelf of the United Kingdom is given as 156 nautical miles in DEP'T OF STATE, BUREAU OF INTELLIGENCE AND RESEARCH, International Boundary Study, Series A, No. 10, Limits in the Sea, Continental Shelf Boundary, The North Sea 7, 9 (1970).
58. *See* OIL & GAS J., June 3, 1974 (insert map of North Sea) for the numerous new fields that have been discovered at distances in excess of 100 nautical miles from shore.

Undesirable Colonial Grab

Some argue that an assertion of United States jurisdiction over the entire United States continental margin would constitute an undesirable colonial "grab." President Johnson, at the commissioning of the Navy research vessel *Oceanographer* on July 13, 1966, stated: "[U]nder no circumstances, we believe, must we ever allow the prospects of rich harvest and mineral wealth to create a new form of colonial competition among the maritime nations. We must be careful to avoid a race to grab and to hold the lands under the high seas. We must ensure that the *deep seas* and the *ocean bottoms* are, and remain, the legacy of all human beings."[59] Seldom has so much leverage been sought to be derived from such a brief rhetorical statement. As Shalowitz has pointed out, the terms *deep seas* and *ocean bottoms* are the key to President Johnson's meaning, and it is grossly unwarranted to treat his statement as a renunciation of broad continental shelf claims by the United States.[60]

The contrary views of the Stratton Commission notwithstanding, this writer has never been able to understand how anyone could charge the United States or any other coastal State with making a colonial grab in asserting a right that the International Court of Justice has characterized as an inherent right, existing ipso facto and ab initio. Moreover, nothing could be clearer than the fact that adoption of the broad definition of Article 1 of the Convention was the handiwork of sometime colonies, not that of former colonial powers. It will be recalled that the White House press release announcing the 1945 Truman Proclamation on the Continental Shelf spoke in terms generally regarded as favorable to a narrow, rather than a broad, shelf.[61] During the subsequent vacillation of the ILC between a broad and a narrow definition, the role of the United States was a passive one.[62] It was not until the Latin American countries pressed for a broad definition at Cuidad Trujillo that the United States affirmatively aligned itself with their views.[63] Even then, it left the fight to

---

59. This passage is quoted, among other places, in 1 NATURAL RESOURCES LAWYER 30 (1968) (emphasis added).
60. Shalowitz, Letter to the Editor, Wash. Post, note 31 *supra.*
61. Presidential Procl., note 13 *supra.* This is as inaccurate as the characterization of the 200-meter isobath as the average depth of the outer edge of the continental shelf. *See* NPC Report, note 5 *supra.* The width of the shelves of the world range from less than 1 km. to as much as 1,200 km. NPC Report, *supra* note 5, at 105.
62. 4 M. WHITEMAN, *supra* note 9, at 833.
63. *Id.* at 837.

them. As Marjorie Whiteman says: "The United States Delegation did not go to the [1958 Geneva] Conference with the intention of sacrificing points of importance on aspects of the law of the sea considered in other committees of the conference in order that there should be a Convention on the Continental Shelf."[64]

Previous mention has been made of the rejection at Geneva in 1958 by a vote of 48 to 20, with 2 abstentions, of a proposal in Plenary Session to strike out the exploitability clause and limit national jurisdiction to the 200-meter isobath. It is highly significant that the developing countries participating in the conference were virtually unanimous in favor of the retention of the exploitability clause and that every former colonial power of Western Europe, with the single exception of Spain abstaining, was among the minority favoring its deletion.[65]

So much for rhetoric as the key to the limits of national jurisdiction over the continental shelf; the significant factor is the passionate regard of the developing countries for what, in United Nations parlance, is euphemistically called "permanent sovereignty over natural resources." It is significant to note that on June 9, 1972, in the Declaration of Santo Domingo, the Latin American countries reiterated the position their spokesmen had taken at Geneva in 1958, calling for their delegations in the United Nations Seabed Committee to promote a study concerning the advisability and timing for the establishment of precise outer limits of the continental shelf, *taking into account the outer limits of the continental rise.*[66]

Future Agreement Dependent on a Narrow Shelf

Some argue that a majority of the delegations at the forthcoming law of the sea conference will favor a narrow shelf and that the others will have to go along in order to obtain an agreement. Robert D. Hodgson and Terry V. McIntyre have prepared an interesting and informative study (Chapter 4) of various types of voting blocs that may emerge at the conference. They point out that the number of possible participants is constantly changing as additional colonies and sheikdoms gain their independence but that, as of February 1971, there were 147 independent States in existence and that one possible grouping of this number, as regards their views on the breadth of the continental shelf, would be into 29 landlocked States with no overseas territories, 22

64. M. Whiteman, *supra* note 18, at 657.
65. *See* U.N. Doc. A/CONF. 13/38 (1958), note 21 *supra.*
66. 11 INT'L LEGAL MATERIALS 892–93 (1972).

shelflocked States with little or no continental margin extending beyond the 200-meter isobath, and 63 States with narrow continental margins, leaving only 33 States with broad continental margins.[67] As United Nations conferences require a two-thirds affirmative vote to open a proposed multilateral treaty for signature, one might readily conclude from these figures that States with a greater interest in a maximum international seabed area and a potential share in the resulting international revenues than in confirmation of coastal State seabed-resource jurisdiction over the entire continental margin could easily control the conference. It would be a blatant error, however, to conclude that their decision would be determinative of the matter.

In his Letter of Transmittal to the Senate of November 22, 1971, urging advice and consent on ratification of the Vienna Convention on the Law of Treaties, President Nixon emphasized the treaty's "strong reaffirmation of the basic principle *pacta sunt servanda*—the rule that treaties are binding on the parties and must be performed in good faith."[68] Article 34 of the Vienna Convention provides that a treaty does not create either obligations or rights for a third State without its consent, and Article 30(4)(b) provides that, when the parties to a later treaty do not include all the parties to an earlier one relating to the same subject matter, as between a State party to both treaties and a State party to only one of them, the treaty to which both States are parties governs their mutual rights and obligations.[69]

It was expressly agreed at Geneva in 1958 that the Geneva Convention should have no expiration clause and that it could not be denounced. In the Plenary Session Andre Gros "explained that the reason for the position taken by the Drafting Committee regarding a denunciation clause was that, to a very large extent, the task of the Conference was to codify customary law; by its nature, such law could not be denounced. Where new law had been made, it had been adopted by general consent, and there would be no point in providing for its denunciation."[70]

It would seem clearly to follow that no party to the Geneva Convention can be deprived of its rights under that Convention without its express consent, despite the apparent efforts of the drafters of the United States Draft Treaty on the International Seabed Area to create universal law.[71] As a consequence, the potential for a viable

---

67. *See* Chapter IV.

68. 11 INT'L LEGAL MATERIALS 234 (1972).

69. 8 INT'L LEGAL MATERIALS 679, 691, 693 (1969).

70. *See* U.N. Doc. A/CONF. 13/38 (1958), *supra* note 21, at 56–57.

71. *See* Jennings, *The United States Draft Treaty on the International Seabed Area— Basic Principles,* 20 INT'L & COMP. L.Q. 433, 435 *et seq.* (1971).

treaty modifying the Convention rests in the hands of the coastal States with significant continental margins rather than those with little or no resources beyond the 200-meter isobath.

The coastline measurements of the world's major political entities are set forth in Table 1. An analysis of this table reveals that seven

Table 1. Coastline measurements of world's major political entities

Order	Political entity	Coastline	% of total	Cum. %
1	USSR	23,098	12.72	12.72
2	Indonesia	19,784	10.89	23.61
3	Australia	15,091	8.31	31.92
4	USA	11,650	6.41	38.34
5	Canada	11,129	6.13	44.47
6	Philippines	6,997	3.85	48.32
7	Mexico	4,848	2.67	50.99
8	Japan	4,842	2.67	53.65
9	Brazil	3,692	2.03	55.69
10	China, Communist	3,492	1.92	57.61
11	Chile	2,882	1.59	59.20
12	UK	2,790	1.54	60.73
13	New Zealand	2,770	1.53	62.26
14	India	2,759	1.52	63.78
15	Italy	2,451	1.35	65.13
16	Madagascar	2,155	1.19	66.31
17	Argentina	2,120	1.17	67.48
18	Spain	2,038	1.12	68.60
19	Turkey	1,921	1.06	69.66
20	Malaysia	1,853	1.02	70.68
21	Cuba	1,747	0.96	71.64
22	Norway	1,650	0.91	72.55
23	Greece	1,645	0.91	73.46
24	Somalia	1,596	0.88	74.34
25	South Africa	1,462	0.81	75.14
26	France	1,373	0.76	75.90
27	*Sweden**	1,359	0.75	76.65
28	Mozambique	1,352	0.74	77.39
29	Saudi Arabia	1,316	0.72	78.12
30	United Arab Republic	1,307	0.72	78.84
31	Thailand	1,299	0.72	79.55
32	Peru	1,258	0.69	80.24
33	Burma	1,230	0.68	80.92
34	Venezuela	1,081	0.60	81.52
35	Iceland	1,080	0.59	82.11
36	Colombia	1,022	0.56	82.67
37	Muscat and Oman	1,005	0.55	83.23

Table 1. Continued

Order	Political entity	Coastline	% of total	Cum. %
38	Iran	990	0.54	83.77
39	Panama	979	0.54	84.31
40	Libya	910	0.50	84.81
41	Morocco	895	0.49	85.30
42	South Vietnam	865	0.48	85.78
43	Angola	806	0.44	86.23
44	Pakistan	750	0.41	86.64
45	South-West Africa	748	0.41	87.05
46	Portugal	743	0.41	87.46
47	*Finland*	735	0.40	87.86
48	South Korea	712	0.39	88.26
49	*Denmark*	686	0.38	88.63
50	Tanzania	669	0.37	89.00
51	Ireland	663	0.37	89.37
52	Southern Yemen	654	0.36	89.73
53	Ceylon	650	0.36	90.09
54	Algeria	596	0.33	90.41
55	Haiti	584	0.32	90.73
56	North Korea	578	0.32	91.05
57	Tunisia	555	0.31	91.36
58	Ethiopia	546	0.30	91.66
59	Spanish Sahara	490	0.27	91.93
60	China, Republic of	470	0.26	92.19
61	Ecuador	458	0.25	92.44
62	Costa Rica	446	0.25	92.69
63	Nicaragua	445	0.24	92.23
64	Yugoslavia	426	0.23	93.17
65	Trucial States	420	0.23	93.40
66	Nigeria	415	0.23	93.63

Source: Sovereignty of the Sea, Geographic Bulletin No. 3 (Revised October 1969), Table II, page 18 *et seq.,* Department of State Publication 7849.

*Italicized countries have no continental margin more than 200 meters beneath the sea. Measurements are according to the criteria set forth in the source. Due to rounding of figures, cumulative percentages do not always match the sums of the individual percentages.

States alone possess 50 percent of all ocean coastlines of the world and that twenty-five States possess 75 percent. There would be enough States to command a two-thirds majority on a narrow redefinition of the continental shelf without the concurrence of a single one of these twenty-five states, but such a treaty would be meaningless. One may go down the list, State after State, beginning with the Soviet Union, whose 23,098 nautical miles of coastline put it at the very top, with precious little indication of support for President Nixon's 1970 proposal that the coastal States renounce all national rights to the seabed resources of the high seas beyond the 200-meter water depth.

Moral Obligations to Relinquish Claims

Some argue that the United States as a great and wealthy nation has a moral obligation to relinquish its claim to seabed resources beyond the 200-meter isobath in favor of the poorer members of the community of nations. There is a tendency on the part of the advocates of a maximum international seabed area to make pejorative remarks about the advocates of United States national jurisdiction over the entire continental margin and to suggest that they are out of step with the greater interests of the United States.[72] The answer to this is that what today may seem to be enlightened action on the part of the United States could, within the space of a single decade, easily prove to have been the height of folly. Even before the president's proposal that the United States and other coastal nations renounce all national rights to the seabed resources of the high seas beyond the water depth of 200 meters and that a substantial portion of the revenues from a suggested trusteeship zone extending from the 200-meter isobath to the outer limit of the continental margin be devoted to international community purposes, the National Petroleum Council had been requested by the Department of the Interior to undertake a comprehensive study of the United States energy outlook to the end of the twentieth century. The reason for this request was Interior's concern that the country was entering a new era of domestic energy shortages.[73] By August of 1970, only ten weeks after the issuance of the President's Statement on United States Oceans Policy, the Director of the Office of Emergency Preparedness found it necessary to reverse his former stand in support of a transition from the existing system of oil import controls to a tariff system and recommend to the President the retention of import controls as an essential national security measure due to the growing danger of inadequate supplies of oil from reasonably secure sources.[74]

72. *See, e.g.*, Henkin, *International Law and "the Interests": the Law of the Seabed*, 63 Am. J. Int'l L. 504 (1969), critically analyzed in Finlay, *The Outer Limit of the Continental Shelf*, note 3 *supra*, and defended in Henkin, *A Reply to Mr. Finlay*, note 52 *supra*.

73. Nat'l Petroleum Council Report, U.S. Energy Outlook (December 1972) 325, Appendix 1.

74. NPC Supplemental Report, *supra* note 5, at 36–37, Appendix C.

On May 3, 1971 the United States Senate showed its concern by the adoption of Senate Resolution 45 calling for a comprehensive study of the energy resources and requirements of the United States and of all related questions relevant to a sound national fuels and energy policy.[75] Finally, on June 4, 1971, the President himself sent a Message to Congress in which he stated:

> For most of our history, a plentiful supply of energy is something the American people have taken very much for granted. In the past twenty years alone, we have been able to double our consumption of energy without exhausting the supply. But the assumption that sufficient energy will always be readily available has been brought sharply into question within the last year. The brownouts that have affected some areas of our country, the possible shortages of fuel that were threatened last fall, the sharp increases in certain fuel prices and our growing awareness of the environmental consequences of energy production have all demonstrated that we cannot take our energy supply for granted any longer.
>
> A sufficient supply of clean energy is essential if we are to sustain healthy economic growth and improve the quality of our national life. I am announcing today a broad range of actions to ensure an adequate supply of clean energy for the years ahead.[76]

In response to the request that had been made of it, the National Petroleum Council submitted an initial appraisal of the United States energy outlook for 1971–85 in July 1971 and a final report in December 1972.[77] The supply and demand of energy are, of course, dependent upon many factors: growth in population and GNP; level of industrial activity; availability and cost of capital funds; and government policies with respect to depletion allowances, investment tax credits, import and price controls, environmental requirements, practices in leasing public lands, and the like. The National Petroleum Council considered the effect of variations in these factors and concluded that under any likely combination of circumstances we were going to be increasingly dependent on imported petroleum and that if the adverse factors influencing domestic supply and demand prevailing at the time of the study were not corrected, by 1985 we could be dependent on oil imports for as much as 65 percent of our oil supplies and 33 percent of our total energy supplies.[78] In the case of natural gas, imports by 1985 might similarly reach the level of 29 per-

75. S. Res. 45, 92d Cong., 1st Sess. (1971).

76. H.R. Doc. No. 92–118, 92d Cong., 1st Sess. (1971).

77. Nat'l Petroleum Council Report, U.S. Energy Outlook: An Initial Appraisal 1971–1975 (July 1971) and Nat'l Petroleum Council Report, U.S. Energy Outlook (December 1972).

78. Nat'l Petroleum Council Report, U.S. Energy Outlook (December 1972) 7.

cent of total gas supply and 5 percent of total energy supplies and would be held to these levels only by the unavailability of additional imports.[79]

These forecasts received all too little public attention at the time they were made, but now the problem has been imprinted on the national consciousness by the events following the outbreak of the "Yom Kippur" War in the Middle East: the Arab boycott of the United States, the cutback in production levels by many of the petroleum exporting countries, and the drastic escalation of oil prices. As a consequence, the quest for increased national self-sufficency in oil and gas is very much the order of the day, though all too many members of Congress seem to be confusing vindictive action against the petroleum industry with attainment of our true national objective.

For technical and economic reasons, the short-range solution will have to be found elsewhere than in the expedited production of oil and gas from the deeper waters of the continental margin. The resources of oil and gas there are enormous, however,[80] and, in the judgment of this writer, the situation just described flashes a bright red warning signal against either the renunciation of our seabed-resource rights beyond the 200-meter isobath or the irrevocable dedication of anything remotely approaching one-half to two-thirds of the government revenues from that portion of the United States continental margin beyond the 200-meter isobath to international community purposes, as the American Draft Treaty would require.[81] In the light of our own

79. *Id.*
80. The United States Geological Survey has estimated total potential resources in place on the United States outer continental shelf between the 200- and 2500-meter isobaths as being between 640 and 800 billion barrels of crude oil, between 50 and 70 billion barrels of natural gas liquids and between 1,590 and 2,230 trillion cubic feet of natural gas. *Hearings on S. Res. 45 Before the Senate Comm. on Interior and Insular Affairs,* 92d Cong., 2d Sess., Ser. 27, pt. 1, at 192, 195 (1972) (article by McKelvey, Wang, Schweinfurth, & Overstreet, Tables 5 & 6). While no part of these resources is listed as recoverable under current economics and technology, this situation is bound to change with the rapid strides in offshore technology and the ever-mounting demand for energy.
81. U.N. Doc. A/AC. 138/22 (1970), 9 INT'L LEGAL MATERIALS 806 (1970) (App. A, 3.1, 4.1, 6.4 & 10.3, App. C, 9.2). Under section 9.2, relatively unimportant fees designed to defray administrative costs of the coastal State are exempt from this sharing obligation. The magnitude of possible future revenues involved may be inferred from the fact that less than 1 percent of the Outer Continental Shelf was under lease as of Dec. 31, 1971, but total U.S. Government revenues from the outer continental shelf through that date were $6,456,688,788. These figures are from the Department of the Interior's reply to a questionnaire of the Senate Committee on Interior and Insular Affairs in connection with *Hearings, supra* note 80, at 114 (Table 52).

critical energy outlook and our inability to help even the most deserving of our less-advantaged foreign friends if we do not keep our own financial house in order, it would seem much the wiser course to maintain our full continental shelf rights under existing international law with the possible exception of the dedications to international community purposes of a modest proportion of the government revenues from the area beyond the 200-meter isobath. Allocations of the magnitude suggested in the United States Draft Treaty should be left to congressional decision on a year-to-year basis.

Conclusion

The writer's best judgment as to what will eventually evolve from the United Nations deliberations is that there will either be a stalemate under which the existing law of the continental shelf will continue in effect unchanged or an agreement will be reached confirming the jurisdiction and control of the coastal States over the seabed resources of either the entire continental margin or a minimum distance from shore (with 200 nautical miles as the leading candidate for the minimum distance), whichever is the greater. The Latin American States seem clearly to want this, if one may judge their wishes in the light of the Declaration of Santo Domingo, and there is good reason to believe that they will carry along with them the great body of developing coastal States facing on the open sea and many of the industrial nations as well.

Other nations have not hastened to join the United States in the Administration's proposal to renounce rights and dedicate a substantial portion of revenues to international community purposes beyond the 200-meter isobath, and their negative reaction should serve as full justification for a substantial retrenchment on our part, if any justification other than our own national interest were needed. In this regard it is gratifying to note that the United States has now expressed its readiness to accept agreement on broad coastal State jurisdiction over seabed resources beyond the territorial sea, conditioned upon suitable recognition being given, as it should be, to the high sea rights of others.[82]

82. *See* Statement of John R. Stevenson, Chairman of the U.S. Delegation to the U.N. Seabed Committee made to Subcommittee II in conjunction with the introduction on July 18, 1973, of the U.S. Draft Articles on the Rights and Duties of States in the Coastal Seabed Economic Area (U.N. Doc. A/AC. 138/SC. II/L. 35, July 16, 1973), Press Release, United States Mission, Geneva, Switzerland, July 18, 1973. *See also* his Letter to the Editor, Wall St. J., Jan., 7, 1974.

Jurisdiction over the Seabed

III

National Seabed Jurisdiction in the Marginal Sea: The South China Sea

Northcutt Ely and J. Michel Marcoux

Introduction: The Limits of National Seabed Jurisdiction as Tested in the South China Sea

One of the problems high on the agenda of the recent debates in the United Nations Seabed Committee, in preparation for the Third United Nations Conference on the Law of the Sea calendared for 1974–75, was the question of the seaward limits of the exclusive jurisdiction of coastal States with respect to the exploration and exploitation of the resources of the seabed and its subsoil adjacent to coastal State land territories. By way of shorthand, we may refer to this problem as one of the limits of national seabed jurisdiction.[1] The jurisdiction referred to is quite special and restricted, in qualitative terms, but quite extensive in geographic, or quantitative, terms.

Qualitatively, the problem is the extent of the coastal State's exclusive competence to control access to, and use of, the land underlying the sea, not the question of its competence to control access to, and use of, the superjacent waters and airspace. Quantitatively, the coastal State's seabed jurisdiction is not limited by the width of the territorial sea but may extend seaward for hundreds of miles beyond it, as, for example, in the North Sea, where the coastal States have licensed petroleum development up to 200 miles seaward of the boundaries they claim for their respective territorial seas. Indeed, the right has no relation, either qualitatively or spatially, to a State's territorial sea jurisdiction. This seabed jurisdiction, which is sui generis, is sometimes referred to as the doctrine of the continental shelf. But this is a misnomer, as will be developed, because the right is not limited by the geological or geomorphic continental shelf.

What, then, are the geographical limits to the exclusive competence of the coastal State to control the exploration and exploitation of the

1. "Seabed jurisdiction" is a more concise phrasing of the longer expression concerning the coastal State's exclusive sovereign rights to the exploration and exploitation of the seabed and subsoil of the submarine areas adjacent to its coast. Convention on the Continental Shelf, *done* Apr. 29, 1958, [1964] 1 U.S.T. 471, T.I.A.S. No. 5578, 499 U.N.T.S. 311, arts. 1, 2, 3 (in force for U.S. June 10, 1964) [hereinafter cited as Continental Shelf Convention].

resources of the seabed that abuts its coast, or, more accurately, abuts its territorial sea? Is there some present restraint in international law, related to depth or distance? Is the key to be found in some test of geological or geomorphic affinity between the seabed and the coastal State's land territories? Is adjacency, in the sense of proximity or contiguity, the key? Is consideration to be given to historic or economic factors? If the seabed is that of a semienclosed or marginal sea, are we dealing with a special case? If the area is adjacent to several States, how shall submarine boundaries be delimited?

These are not academic questions but urgent practical ones. The continental margins hold the most promising untapped petroleum potential in the world. Private sources confirm that offshore of Indonesia a well has been drilled successfully in water over 2,000 feet deep. The National Petroleum Council reported in 1969 that in its judgment "[w]ithin less than five years, technology will allow drilling and exploitation in water depths up to 1,500 feet (457 meters). Within ten years technical capability to drill and produce in water depths of 4,000–6,000 feet (1,219–1,829 meters) will probably be attained."[2] It is essential to know whether some coastal State has authority to license the development of a particular area or whether no State has present competence to do so, thus postponing resource development until the community of nations comes to some agreement with respect to the seaward limits of coastal States' seabed jurisdiction or to agreement on some form of international regime that will have competence to license development of the area in question.

The writers will attempt to supply answers and to test them against the formidable challenge offered by the locus of the South China Sea. Its high rating as a potential source of petroleum[3] makes it a particu-

---

2. UNITED STATES NATIONAL PETROLEUM COUNCIL, PETROLEUM RESOURCES UNDER THE OCEAN FLOOR 8 (1969).

3. Emery & Ben-Avraham, *Structure and Stratigraphy of China Basin,* 56 AM. ASS'N PETROLEUM GEOLOGISTS BULL. 839, 845 (1972) [hereinafter cited as Emery]. *See also* Mainguy, *Regional Geology and Petroleum Prospects of the Marine Shelves of Eastern Asia,* U.N. ECON. COMM'N FOR ASIA & THE FAR EAST, COMM. FOR COORDINATION OF JOINT PROSPECTING FOR MINERAL RESOURCES IN ASIAN OFFSHORE AREAS, 3 TECHNICAL BULL. 91, 103 (1970). For statistics on offshore resources, see ALBERS, *et al.,* SUMMARY PETROLEUM AND SELECTED MINERAL STATISTICS FOR 120 COUNTRIES, INCLUDING OFFSHORE AREAS (1973, U.S. Geological Survey Professional Paper 817). For an analysis of governmental controls on pricing and production from the seabed, see Marcoux, *Seabed Mineral Resource Production and the Free Market,* 6 NATURAL RESOURCES LAWYER 217 (1973). For an overview of Asian resource potential, see THE OFFSHORE HYDROCARBON POTENTIAL OF EAST ASIA, A REVIEW OF INVESTIGATIONS, 1966–1973, Office of the Project Manager/ Co-Ordinator, UNDP Tech. Support for Regional Offshore Prospecting in East Asia (Bangkok, Feb. 1974).

larly appropriate testing ground, assuming that complicated jurisdictional and boundary problems can be solved.

The Rationale of National Jurisdiction over Seabed Resources

The Historical Background of the Continental Shelf Doctrine

The concept of coastal State jurisdiction over seabed resources seaward of its territorial sea is a relatively new one in international law. Its growth has been explosive, responding to the necessities for a rule of law as petroleum exploration has extended into progressively deeper waters. Thus, in 1952 a distinguished British arbitrator, Lord Asquith of Bishopstone, held in the Abu Dhabi arbitration award that as of 1939 there had been no "doctrine of the continental shelf."[4] His conclusion may have been dubious. Many States have long exercised exclusive competence over certain types of seabed resources, such as pearls and sponges, in waters that are adjacent to their coasts but well beyond the limits of the territorial sea that they assert.[5]

The modern concept of a continental shelf was first articulated clearly in President Truman's Executive Proclamation of 1945, which stated that "the continental shelf may be regarded as an extension of the land-mass of the coastal nation and thus naturally appurtenant to it" and that "the United States regards the natural resources of the subsoil and sea bed of the continental shelf beneath the high seas but contiguous to the coasts of the United States as appertaining to the United States, subject to its jurisdiction and control. . . ."[6] The Truman Proclamation is notable as being the recognized beginning of both the conventional and customary law that has evolved rapidly since that date: the former as stated in the 1958 Geneva Convention on the Continental Shelf, the latter from the practice of States, and as articulated in the 1969 Judgment of the International Court of Justice in the *North Sea Continental Shelf Cases.*[7] Account must be taken of a third potential influence on the law of this subject: the proceedings of the United Nations in preparation for the Third United Nations Conference on the Law of the Sea convened June 20, 1974, in Caracas, Venezuela. Despite deadlock on August 29, 1974, there was strong support among the 148 participating nations for a *lex ferenda*

---

4. In the Matter of an Arbitration between Petroleum Development (Trucial Coast) Ltd. and the Sheikh of Abu Dhabi, 1 INT'L & COMP. L.Q. 247, 258 (1952).
5. I. BROWNLIE, PRINCIPLES OF PUBLIC INTERNATIONAL LAW 214–15 (1973).
6. 10 Fed. Reg. 12303 (1945), 13 DEP'T STATE BULL. 485 (1945).
7. North Sea Continental Shelf Cases, [1969] I.C.J. 3 (decided Feb. 20, 1969), *reprinted in* 8 INT'L LEGAL MATERIALS 340 (1969).

doctrine of a national marine economic resource zone, or patrimonial sea, that would tend to extend, rather than constrict, the geographical extent of the coastal State's seabed jurisdiction.

The 1958 Convention on the Continental Shelf requires attention, even in construing customary law, because of the *North Sea Continental Shelf Cases* decision, which held that Articles 1, 2, and 3 stated customary as well as conventional law.

Article 1 states: "For the purpose of these articles, the term 'continental shelf' is used as referring (a) to the seabed and subsoil of the submarine areas adjacent to the coast but outside the area of the territorial sea, to a depth of 200 metres or, beyond that limit, to where the depth of the superjacent waters admits of the exploitation of the natural resources of the said areas; (b) to the seabed and subsoil of similar submarine areas adjacent to the coasts of islands." Continuing in Article 2, the Convention declares:

1. The coastal State exercises over the continental shelf sovereign rights for the purpose of exploring it and exploiting its natural resources.

2. The rights referred to in paragraph 1 of this article are exclusive in the sense that if the coastal State does not explore the continental shelf or exploit its natural resources, no one may undertake these activities, or make a claim to the continental shelf, without the express consent of the coastal State.

3. The rights of the coastal State over the continental shelf do not depend on occupation, effective or notional, or on any express proclamation.

These provisions must be read in conjunction with Article 3 which provides: "The rights of the coastal State over the continental shelf do not affect the legal status of the superjacent waters as high seas, or that of the airspace above those waters." The Convention, particularly Articles 1 and 2, which are of special importance here, has a noteworthy history.

In 1949, after more than a score of States had issued proclamations patterned on the Truman Proclamation,[8] the United Nations General Assembly instructed the International Law Commission to undertake preparation for a conference on the Law of the Sea. The Commission released three successive drafts of treaties, with commentaries, in 1951, 1953, and 1956. On the question of seabed jurisdiction, these drafts showed an interesting evolution.

In 1951 the Commission published a proposed Draft Treaty on the Law of the Sea, which provided that the coastal State should have exclusive seabed jurisdiction in all adjacent submarine areas capable of

8. For the texts of national claims to the continental shelf between 1945 and 1949 by the United States, Mexico, Panama, Argentina, Chile, Peru, Nicaragua, Costa Rica, and Saudi Arabia, see 46 UNITED STATES NAVAL WAR COLLEGE, INTERNATIONAL LAW DOCUMENTS 1948–1949, at 182–96 (1950).

exploitation.[9] In 1953 the Commission, in a new draft, reversed itself on this point, proposing the 200-meter isobath as a limit to coastal State jurisdiction instead.[10] This new limit was unacceptable to the American States. In 1956 twenty American States met at Ciudad Trujillo and unanimously recommended the addition of the criterion of exploitability, regardless of depth, limited only by the criterion of adjacency. The language they proposed read: "The sea-bed and subsoil of the continental shelf, *continental and insular terrace,* or other submarine areas, adjacent to the coastal state, outside the area of the territorial sea, and to a depth of 200 meters or, beyond that limit, to where the depth of the superjacent waters admits of the exploitation of the natural resources of the sea-bed and subsoil, appertain exclusively to that state and are subject to its jurisdiction and control."[11] The term "continental terrace" was defined as "that part of the submerged land mass that forms the shelf and the slope" and the term "continental slope" or "inclination" as "the slope from the edge of the shelf *to the greatest depths.*"[12] The Conference concluded that "[t]he American States are especially interested in utilizing and conserving the existing natural resources on the American terrace (shelf and slope)" and that "[t]he utilization of the resources of the shelf cannot be technically limited, and for this reason the exploitation of the continental terrace should be included as a possibility in the declaration of rights of the American States."[13] This position was presented to, and accepted by, the International Law Commission. In 1956 it released its third draft, which provided, on this point:

Article 67

For the purposes of these articles, the term "continental shelf" is used as referring to the seabed and subsoil of the submarine areas adjacent to the coast but outside the area of the territorial sea, to a depth of 200 metres (approximately 100 fathoms), or, beyond that limit, to where the depth of the superjacent waters admits of the exploitation of the natural resources of the said areas.

Article 68

The coastal State exercises over the continental shelf sovereign rights for the purpose of exploring and exploiting its natural resources.[14]

9. [1951] 2 Y.B. INT'L L. COMM'N 141.
10. [1953] 2 Y.B. INT'L L. COMM'N 212.
11. Inter-American Specialized Conference on "Conservation of Natural Resources: The Continental Shelf and Marine Waters," *Final Act,* Ciudad Trujillo, March 15–28, 1956, ORGANIZATION OF AMERICAN STATES CONFERENCES AND ORGANIZATIONS SERIES NO. 50, at 13 (1956) (emphasis added).
12. *Id.* at Doc. 90 (emphasis added).
13. Parentheses in original.
14. [1956] 2 Y.B. INT'L L. COMM'N 253, 296–97.

The Commission's report contained the following explanation:

(4) At its eighth session, the Commission reconsidered this provision. It noted that the Inter-American Specialized Conference on "Conservation of Natural Resources: Continental Shelf and Oceanic Waters", held at Ciudad Trujillo (Dominican Republic) in March 1956, had arrived at the conclusion that the right of the coastal State should be extended beyond the limit of 200 metres, "to where the depth of the superjacent waters admits of the exploitation of the natural resources of the seabed and subsoil". Certain members thought that the article adopted in 1953 should be modified. . . . Other members contested the usefulness of the addition, which in their opinion unjustifiably and dangerously impaired the stability of the limit adopted. The majority of the Commission nevertheless decided in favour of the addition.

(5) The sense in which the term "continental shelf" is used departs to some extent from the geological concept of the term. The varied use of the term by scientists is in itself an obstacle to the adoption of the geological concept as a basis for legal regulation of this problem. . . .

(9) . . . The Commission considered that some departure from the geological meaning of the term "continental shelf" was justified, provided that the meaning of the term for the purpose of these articles was clearly defined. It has stated this meaning of the term in the present article.[15]

In 1958, the plenary Law of the Sea Conference considered the Commission's 1956 draft, amended it slightly, and adopted it in the form now appearing in Article 1 of the Convention.[16]

During the course of the debate, it became clear that the Conference was adopting the Ciudad Trujillo concept as a compromise.[17] Amendments were rejected that would have deleted the exploitability criterion and thus would have restricted the coastal State's jurisdiction to the 200-meter isobath.[18] After adopting the language now appearing in the Convention, the Conference likewise rejected an amendment that would have related jurisdiction solely to exploitability.[19] There is no question that the effect of the language of Article 1, as adopted, was intended to be exactly that proposed by the 1956 Declaration of Ciudad Trujillo. For example, the chief American ne-

15. *Id.*

16. The International Law Commission had held many meetings over several years, and Articles 67–73 of its 1956 Draft on the Law of the Sea "formed the basic working paper of the Geneva Conference." Young, *The Geneva Convention on the Continental Shelf: A First Impression,* 52 Am. J. Int'l L. 733, 734 (1958). The ILC paper is found at [1956] 2 Y.B. Int'l L. Comm'n 253, 295–301, 11 U.N. GAOR Supp. 9, at 40–45, U.N. Doc A/3159 (1956) *reprinted in* 51 Am. J. Int'l L. 154, 242 (1957).

17. *See* note 15 *supra.*

18. U.N. Doc. A/CONF. 13/C. 4/L. 8 (1958) (Lebanon).

19. U.N. Doc. A/CONF. 13/C. 4/L. 11 (1958) (Korea).

gotiator reported to the United States Senate: "The clause which protects the right to utilize advances in technology at greater depths beneath the oceans was supported by the United States and was in keeping with the inter-American conclusions at Ciudad Trujillo in 1956. It was included in the ILC 1956 draft."[20]

The net result was summed up in the 1968 Interim Report of the American Branch of the International Law Association:

> The jurisdiction of coastal states with respect to the natural resources of the sea-bed and sub-soil areas under the high seas is determined by the Geneva Convention on the Continental Shelf. By that instrument the community of nations has decided that the interests of mankind are best served by reserving to coastal states exclusive sovereign rights in the natural resources of the sea-bed and sub-soil of the submarine areas adjacent to their coasts, not only to the 200 meter depth, but beyond that depth "to where the depth of the super-jacent water admits of the exploitation of the natural resources." The basis for this recognition of exclusive mineral jurisdiction is twofold: the predominant interest of the coastal state in the bed of the sea adjacent to its shores, and the necessity for certainty as to what law is applicable to that sea-bed. To date, some three-score nations have given recognition to the principles of that Convention, 36 by ratifying it, the others by adopting major provisions of it in domestic legislation or regional agreements. From the wide acceptance of the principles set forth in the Convention, even by states which are not parties, it is clear that they constitute part of customary international law.
>
> . . . .
>
> As a general rule, the limit of adjacency may reasonably be regarded as coinciding with the foot of the submerged portion of the continental land mass. There is strong support for this view in the drafting history of the Convention [citations omitted], although other interpretations have been advanced.[21]

The same Committee's 1970 Interim Report said: "[T]he Committee stands on its prior position that rights under the 1958 Geneva Convention on the Continental Shelf extend to the limit of exploitability

---

20. *Hearing on Conventions on the Law of the Sea Before the Senate Comm. on Foreign Relations,* 86th Cong., 2d Sess. 108 (1960).

21. INTERNATIONAL LAW ASSOCIATION, INTERIM REPORT OF THE AMERICAN BRANCH COMM. ON DEEP SEA MINERAL RESOURCES IX–X (1968). As of Jan. 1, 1974, 52 nations had ratified or acceded to the Convention, an increase of 16 in the six-year interval after the writing of the 1968 Report, U.S. DEP'T. OF STATE, TREATIES IN FORCE 354–55 (Jan. 1, 1974, Pub. 8755). The statement that the Convention principles constitute customary international law was prophetic, being echoed in the decision of the International Court of Justice the following year in the North Sea Continental Shelf Cases, [1969] I.C.J. 3.

existing at any given time, within an ultimate limit of adjacency which would encompass the entire continental margin."[22]

This view of the American Branch reports was adopted by the Special Subcommittee on Outer Continental Shelf of the Senate Committee on Interior and Insular Affairs which said in a report published in 1971:

> Adjacency as applied to the legal Continental Shelf means the seaward limit of the natural prolongation of the submerged land continent. The submerged land continent encompasses the geomorphic Shelf, slope and rise. Thus, the rule laid down in the 1958 Geneva Convention on the Continental Shelf, as we interpret it, holds that the sovereign rights of coastal nations to explore and exploit their legal Continental Shelf extend to the limit of exploitability existing at any given time within an ultimate limit of adjacency which encompasses the entire continental margin.[23]

The Subcommittee concluded:

> We construe the heart of our sovereign rights under the 1958 Geneva Convention to consist of the following:
> (1) The exclusive ownership of the mineral estate and sedentary species of the entire continental margin;
> (2) The exclusive right to control access for exploration and exploitation of the entire continental margin; and
> (3) The exclusive jurisdiction to fully regulate and control the exploration and exploitation of the natural resources of the entire continental margin.[24]

Substantially similar opinions were expressed in the 1968 Report of the British Branch of the International Law Association,[25] and in the 1972 Report of the Australian Branch.[26] Moreover, the Convention on the Continental Shelf has now been ratified by fifty-two States. At

22. INTERNATIONAL LAW ASSOCIATION, INTERIM REPORT OF THE AMERICAN BRANCH COMM. ON DEEP SEA MINERAL RESOURCES 1 (1970).
23. SENATE COMM. ON INTERIOR & INSULAR AFFAIRS, 91ST CONG., 2D SESS., REPORT ON OUTER CONTINENTAL SHELF 16 (Comm. Print 1971).
24. *Id.* at 29 (footnote omitted). In these writers' view, the coastal State's jurisdiction to control the drilling of exploratory wildcat wells is geographically coextensive with its jurisdiction with respect to production if the wildcat is successful. The spudding-in of a well capable of penetrating formations at depth, with the intent and capability of producing the oil therefrom, constitutes exploitation, and the coastal State's right to permit or prohibit that operation exists *ab initio,* and is not contingent, retroactively, on the well's success in obtaining commercial production. Note the provisions of Article 5(8) which require the coastal State's consent for scientific research on the continental shelf.
25. INTERNATIONAL LAW ASSOCIATION, REPORT OF THE BRITISH BRANCH COMMITTEE ON DEEP SEA MINING (1968).
26. INTERNATIONAL LAW ASSOCIATION, SUPPLEMENT TO THE 1970 REPORT OF THE AUSTRALIAN BRANCH COMMITTEE ON DEEP SEA MINING (1972).

least fifty coastal States have granted leases and licenses in areas beneath waters of a greater depth than 200 meters, evidencing a practical construction of both conventional and customary law.

Summary: Coastal State Seabed Jurisdiction
Under the Convention on the Continental Shelf

Under the Convention, the coastal State's seabed jurisdiction is controlled by two criteria, adjacency and exploitability.

As to adjacency, in view of the legislative history of the Convention, the area encompassed by the criterion of adjacency is coextensive with the continental terrace, that is, the shelf and slope. The seaward edge of the slope is bounded by the junction of the continental crust with the rocks of the abyssal ocean floor. Where this junction is overlaid by the debris that constitutes the continental rise, the coastal State's ultimate seabed jurisdiction, under the Convention, thus encompasses the portion of the rise that is landward of that underlying junction. As will be developed, there is no substantive difference between this concept of adjacency as fixing the boundary of the coastal State's seabed jurisdiction under the Convention and the concept of the prolongation of land territories articulated by the International Court of Justice in the *North Sea Continental Shelf Cases*.

As to the criterion of exploitability, under the Convention on the Continental Shelf coastal State jurisdiction is coextensive with "exploitability," limited, however, to the seaward extent of the submerged continental landmass by the notion of "adjacency." Thus it is necessary to determine the parameters of the "exploitability" criterion. Some commentators have argued that coastal State jurisdiction progresses seaward from the 200-meter isobath in a stepwise manner as exploitability is demonstrated at ever increasing depths. Others, with whom these writers join, point out that because demonstrated exploitability automatically brings the seabed area in question within the Convention's definition of "continental shelf" regardless of the nationality of the operator, and because the Convention expressly provides in Article 2 that no one may undertake exploration or exploitation of natural resources on the continental shelf, or, indeed, undertake scientific research there (Article 5) without the consent of the coastal State, coastal State jurisdiction for all practical purposes embraces the entire continental margin, regardless of the depth of demonstrated exploitability. The power to exclude others is the ultimate test of jurisdiction and this power is not conditioned on exploitability. The distinction drawn by these two points of view may now be moot, since recent deep sea mining operations have demonstrated a present commercial capability to recover manganese nodules from areas of the deep ocean floor well beyond the maximum possible limits of coastal State jurisdiction. Thus, unless a distinction can be made between hard mineral exploitability and petroleum exploitability—a distinction for which there is no basis in the Convention itself—coastal State jurisdiction (to the extent that it is controlled by the exploitability criterion) has already been extended to the maximum limits allowed by the adjacency criterion of the Convention on the Continental Shelf, *viz.,* to the seaward extent of the submerged continental landmass.

Customary Law and the Concept of the Prolongation of Land Territories

The International Court of Justice in the *North Sea Continental Shelf Cases* determined the principles that should control negotiations as to the lateral seabed boundaries between Germany and Denmark and between Germany and the Netherlands under customary international law. In doing so, it articulated a natural prolongation test for determining the geographical extent of the coastal State's seabed jurisdiction. The Court held, moreover, that Article 2 of the Convention was an exposition of customary as well as conventional law. Indeed, as Professor R. Y. Jennings has pointed out, the Court deemed the principle of prolongation of the continental landmass to have been "enshrined" in Article 2.[27]

With respect to Germany's claim to a "just and equitable share" of the seabed,[28] the Court said:

Delimitation in an equitable manner is one thing, but not the same thing as awarding a just and equitable share of a previously undelimited area, even though in a number of cases the results may be comparable, or even identical.

More important is the fact that the doctrine of the just and equitable share appears to be wholly at variance with what the Court entertains no doubt is the most fundamental of all the rules of law relating to the continental shelf, enshrined in Article 2 of the 1958 Geneva Convention, though quite independent of it,—namely that the rights of the coastal State in respect of the area of continental shelf that constitutes a natural prolongation of its land territory into and under the sea exist *ipso facto* and *ab initio,* by virtue of its sovereignty

27. Jennings, *The Limits of Continental Shelf Jurisdiction: Some Possible Implications of the North Sea Case Judgment,* 18 INT'L & COMP. L.Q. 819, 822 (1969).
28. North Sea Continental Shelf Cases, I.C.J. Pleadings, Oral Arguments, Documents, at 30–36 (1968) (Vol. I).

over the land, and as an extension of it in an exercise of sovereign rights for the purpose of exploring the seabed and exploiting its natural resources. In short, there is here an inherent right. In order to exercise it, no special legal process has to be gone through, nor have any special legal acts to be performed. Its existence can be declared (and many States have done this) but does not need to be constituted. Furthermore, the right does not depend on its being exercised. To echo the language of the Geneva Convention, it is "exclusive" in the sense that if the coastal State does not choose to explore or exploit the areas of shelf appertaining to it, that is its own affair, but no one else may do so without its express consent.[29]

The writers note the expression "natural prolongation of its land territory," a concept that reappears throughout the Opinion, and the Court's appraisal of Article 2 of the Convention as "enshrining" this basis of jurisdiction, which of course is quite independent of proof of exploitability. Another paragraph of the Court's Opinion expands this concept:

More fundamental than the notion of proximity appears to be the principle—constantly relied upon by all the Parties—of the natural prolongation or continuation of the land territory or domain, or land sovereignty of the coastal State, into and under the high seas, via the bed of its territorial sea which is under the full sovereignty of that State. There are various ways of formulating this principle, but the underlying idea, namely of an extension of something already possessed, is the same, and it is this idea of extension which is, in the Court's opinion, determinant. Submarine areas do not really appertain to the coastal State because—or not only because— they are near it. They are near it of course; but this would not suffice to confer title, any more than, according to a well-established principle of law recognized by both sides in the present case, mere proximity confers *per se* title to land territory. What confers the *ipso jure* title which international law attributes to the coastal State in respect of its continental shelf, is the fact that the submarine areas concerned may be deemed to be actually part of the territory over which the coastal State already has dominion,—in the sense that, although covered with water, they are a prolongation or continuation of that territory, an extension of it under the sea. From this it would follow that whenever a given submarine area does not constitute a natural—or the most natural—extension of the land territory of a coastal State, even though that area may be closer to it than it is to the territory of any other State, it cannot be regarded as appertaining to that State;—or at least it cannot be so regarded in the face of a competing claim by a State of whose land territory the submarine area concerned is to be regarded as a natural extension, even if it is less close to it.[30]

Applying this principle, the Court held that the seabed area lying be-

29. [1969] I.C.J. 3, 22 (paras. 18–19).
30. *Id.* at 31 (para. 43).

tween the coast of Germany and that of England was the common prolongation of the land territories of those two nations, and that Germany was entitled to recognition of her interests in this area. The delimitation finally agreed upon by Denmark, Germany, and the Netherlands, on the basis of principles articulated by the Court in the North Sea cases, recognized Germany's interest out to the Germany/England median line, some 194 miles out from the German coast.

How far to sea does the prolongation of territory concept apply in deciding whether a given coastal State has exclusive competence to license exploration and exploitation? Does the geomorphic continental shelf set the limit, or does geology? That is, does the topography of the submarine area control, or does its structure? Or does the coastal nation's seabed jurisdiction extend down the continental slope to its base, past the continental shelf, as the Ciudad Trujillo Conference insisted? If the seaward slope of the continental shelf is not the same thing as the seaward slope of the continental margin because of intervening troughs or islands, what principles control the demarcation of seabed jurisdiction in this instance?

First, it appears desirable to establish the relationship among some of these terms. The report of the National Petroleum Council gives this excellent summary of the issues involved:

The surface of the earth, whether [or not] water covered, can be divided generally into two fundamentally distinct geomorphic units—the *ocean basins* and the *continental platforms*. These units are the surficial reflection of fundamental lateral geological differences in the character of the earth's crust down to the mantle. The continental crust differs from the oceanic crust in (1) higher surface elevation, (2) greater depth to mantle, (3) greater thickness, (4) lesser density, (5) lower seismic velocity, and (6) more acidic rock composition.

The change from continental to oceanic crust is the most distinctive recognizable lateral change in the earth's lithologic character. Viewed on a world scale it is impressively sharp. This is true even though locally and in detail the change often appears to take place through a transition zone or indeterminate zone many kilometers or even hundreds of kilometers wide, and many of the distinguishing features are determinable in detail only approximately and only by means of subsurface geophysical measurements.

The principal reflection at the surface of the fundamental change from continental to oceanic crust is the topographic scarp known as the continental slope, resulting from the isostatic effect tending to make the continental masses stand high relative to the adjacent oceanic areas.

The continental slope has been termed "by far the steepest, longest, and highest topographic feature on the earth's surface." The continental slope is the frontal edge of the submerged continent; and the base of the continental slope, coinciding approximately with the outer limits of the continental crust, constitutes the most distinct, the most profoundly significant, and the only natural surface feature which can be used as a guideline to the outer limits of the continent. Hence, it is a logical starting point for localizing the approximate outer limit of coastal-state jurisdiction.

However, even the base of the slope cannot be defined sharply enough to serve as a boundary of mineral resources jurisdiction in itself, and it should be used only as a guideline to a more precise boundary. Moreover, it should be emphasized that where continental rises are developed adjacent to the continental slope, the sediments of these rises will overlap the lower part of the slope so that the true boundary marking the outer limits of the continental block must be drawn to include not only the slope but also the landward portion of the rise.[31]

The 1968 Interim Report of the Committee on Deep Sea Mineral Resources of the American Branch of the International Law Association called attention to the necessity that the boundary of national seabed jurisdiction bear a logical relation to the physical facts:

From the geological standpoint, this interface at the submerged continental margin is a profound natural boundary. Characterized by a marked change of structure between the continental mass and the crust of the deep ocean basins, it is generally to be found at a depth of from 2,000 to 3,000 meters. As stated recently by the United States representative in the Technical and Economic Working Group of the United Nations Ad Hoc Committee:

The composition of the continents, including their submerged parts, is basically different from that of the oceanic crust of the deep ocean basins. The boundary between the two is one of the most profound natural interfaces. It is gradational in many places and not easily established by direct observation, but generally occurs near the base of the continental slopes at a depth of about 2500 meters.

. . . The gradational interface between the submerged parts of the continents and the ocean basins naturally fixes the seaward limit of any continental feature, and is from the scientific point of view the conceptual boundary between continental and oceanic seabed resources. It is important to recognize, however, that neither this nor any other geologic or topographic boundary is sufficiently distinct and consistent to serve by itself as the means of defining a precise juridical boundary.[32]

As to the juridical boundary, an eminent petroleum geologist has concluded:

[A] consideration of all legal, political, economic and scientific realities, combine to make the base of the continental slope the optimum general guide to the outer boundary of exclusive coastal state jurisdiction over ocean bot-

31. NPC, *supra* note 2, at 10.
32. INTERNATIONAL LAW ASSOCIATION, INTERIM REPORT OF THE AMERICAN BRANCH COMMITTEE ON DEEP SEA MINERAL RESOURCES X (1968).

tom resources. Such a boundary gives recognition not only to the natural oceanward extension of the domain of each coastal nation, but also to the inclusion under its jurisdiction of the bottom resources over which it is the entity most practicably suited to exercise control, and over which many coastal nations (including the U.S.) have already made such commitments.[33]

The criterion of adjacency is thus the counterpart of the Court's criterion of prolongation of land territories.[34] In other words, the Convention's criterion of adjacency and the Court's matching criterion of prolongation are answers to the false fear that the exploitability criterion of the Convention might carry coastal States' jurisdiction to the middle of the Atlantic and Pacific oceans, converting them into exclusively national jurisdictional lakes. That particular worry is a frequently repainted red herring.

33. Speech of Hollis D. Hedberg, presented at the Annual Meeting of the American Association of Petroleum Geologists, Dallas, Texas (Apr. 15, 1969).
34. The American Bar Association has adopted a resolution on the concept of the prolongation of land territories:

"NOW, THEREFORE, BE IT RESOLVED, That the American Bar Association:

"*As to Seabed Resources of the Continental Margin*

"(1) REITERATES its position 'that within the area of exclusive sovereign rights adjacent to the United States, the interests of the United States in the natural resources of the submarine areas be protected to the full extent permitted by the 1958 Convention on the Continental Shelf,' and asserts that these areas encompass or with advancing technology will encompass the full extent of the continental margin adjacent to the United States. The environment must be adequately protected, and other uses of the ocean must be accommodated. Similar rights and obligations are to be recognized in all other coastal States. If an 'economic resource zone' is to be agreed upon, in which the coastal State shall have exclusive rights to seabed resources, the proposed width of 200 nautical miles is acceptable, provided that the exclusive seabed jurisdiction of the United States should be protected to that distance or to the full width of the continental margin, whichever is greater at any given point on the coast. Any treaty commitment for contributions of governmental revenues from the American continental margin for international community purposes should be limited in amount, any larger contributions being reserved for appropriation by Congress in the light of the overall national interest from year to year.

"(2) SUPPORTS the view that the portions of the U.S. Outer Continental Shelf in waters deeper than 200 meters, being now clearly within the exclusive resources jurisdiction of the United States, acting through the Congress, should remain so, and their subjection to any future international treaty should be limited to standards for the prevention of unreasonable interference with other uses of the ocean, for the protection of the ocean from pollution, for the protection of the integrity of investments, and for the compulsory settlement of disputes." American Bar Association Resolution on Natural Resources of the Sea, adopted by the House of Delegates on Aug. 6, 1973, ABA, Section and Committee Reports to the House of Delegates, Report 115(C) (1973), *reproduced in* 6 NATURAL RESOURCES LAWYER 589, 590 (1973).

The Particular Case of Marginal Seas

The term *marginal sea* as applied by many writers describes seas enclosed by island arcs,[35] in the way that the East China Sea is enclosed by the Ryukyu Islands chain, or the South China Sea is enclosed by the island arcs of the Philippines and Indonesia, or the Caribbean Sea is enclosed by the island arc of the Antilles. Geographers identify some forty marginal or semienclosed seas lying along the coasts of the continents. Although they occupy only about 1 percent of the area of the world's oceans, they are estimated to contain some 17 percent of the sediments in all the seabeds of the world.[36] Since petroleum is found only in sedimentary formations, the marginal seas, rather than the deep oceans, will be the primary targets for petroleum exploration as the industry's capability for deep water operations increases. These marginal seas, then, are the true areas where the conflict between various contentions as to the limits of national jurisdiction is relevant. Since the continental margin is considered to terminate in a trench or trough seaward of the island arc that encloses the sea in question, the term "marginal sea" is an apt one. The question as to whether the whole of the enclosed seabed is identifiable geologically with the continental crust, or whether, in some instances, it is identifiable with the oceanic crust that lies beyond the island arc and its flanking trench, or whether in some cases these marginal seas should be considered as a third classification, sui generis,[37] cannot affect the political realities. These realities are that the mainland and island States whose territories collectively constitute the boundaries of the surface of the semienclosed sea have a special interest in that sea, and operations there are of special concern to those States, as indicated in the Truman Proclamation of 1945.[38]

35. Karig, *Structural History of the Mariana Island Arc System,* 82 GEOLOGICAL SOC'Y AMERICA BULL. 323 (1971); Katsumata & Sykes, *Seismicity and Tectonics of the Western Pacific: Izu-Marianna-Caroline and Ryukyu-Taiwan Regions,* 74 J. GEOPHYSICAL RESEARCH 5923 (1969); Beloussov & Ruditch, *Island Arcs in the Development of the Earth's Structure (Especially in the region of Japan and the Sea of Okhotsk),* 69 J. GEOLOGY 647 (1961); Koto, *Geological Structure of the Ryukyu Arc,* 5 J. GEOLOGICAL SOC'Y TOKYO 1 (1897) (trans. K. Musya, U.S. Geological Survey, Tokyo, Feb. 7, 1950).

36. Menard, *Transitional Types of Crust Under Small Ocean Basins,* 72 J. GEOPHYSICAL RESEARCH 3061, 3067 (1967): "[a]bout one-sixth of all identifiable ocean basin sediment is in small basins with only about 1% of the total oceanic area. The total volume of sediment in oceanic trenches, including those almost filled, is relatively trivial. The greatest volumes of thick sediment accumulating anywhere in the world are in small ocean basins."

37. Burk, *Global Tectonics and World Resources,* 56 AM. ASS'N PETROLEUM GEOLOGISTS BULL. 196, 197 (1972).

38. *See* note 6 *supra.*

The 1969 report of the National Petroleum Council took special note of this problem:

> To mention one last special case, in semienclosed seas the surrounding coastal nations may well have a justifiable claim to natural resources jurisdiction to the median line, even where that line lies beyond the outer edge of a typical continental margin. Though the matter does not appear to have been discussed in the preparatory works leading to the choice of language employed in Article 1 of the Convention, the words actually used would appear broad enough to sustain such a claim in the case of any semienclosed sea, such as the North Sea and the Gulf of Mexico, where the conditions of appurtenance or proximity are such as to meet the test of adjacency set forth in Article 1.[39]

The present writers would not wish, however, to condition a legal conclusion as to the seabed jurisdiction of a coastal State merely on proof that the drill will demonstrate that the geology of the seabed is identical to that of the formations of the State's land territories. Security of tenure requires that the competence of the grantor State be known before, rather than after, the concessionaire has drilled. In our view, the law does not compel a conclusion against good sense in this respect. To the contrary, the concept of appurtenance, of the relation of the State's land territory to the State's competence to govern the resources of the seabed, is satisfied simply by a demonstration that the area in question is a semienclosed sea bounded by that State, as distinguished from the open ocean. The doctrine of the semienclosed sea is a geographical cognate of the geological rationale of prolongation, both being articulations of the basic legal concept of appurtenance: the thought that the shore dominates the seabed that adjoins it. The writers would apply the concept of appurtenance to the whole continental margin, including marginal seas and all island arcs that enclose those seas, on a geomorphic or geographical rationale, and not on a criterion of geologic identity. The concept that geology must not be the ultimate criterion for demarcation of the limits of national seabed jurisdiction is consistent with recent proposals for a 200-mile patrimonial sea, or coastal marine resource zone governed by the adjacent State.

The "Patrimonial Sea"

At the 1971 summer meeting of the United Nations Seabed Committee, the Venezuelan representative, Dr. Aguilar, suggested that a good compromise agreement might envisage the following marine spaces:

39. NPC, *supra* note 2, at 57.

(1) A territorial sea under the coastal State's exclusive sovereignty and jurisdiction, with a reasonable width of, say, twelve miles;

(2) An economic zone, called the patrimonial sea, not more than 200 miles in width from the base line of the territorial sea. In that zone, there would be freedom of navigation and overflight but the coastal State would have an exclusive right to all resources;

(3) That part of the continental shelf not covered by the patrimonial sea which would extend to a depth not exceeding 200 metres and over which the various States concerned would maintain their existing rights.[40]

The Venezuelan proposal attracted wide support, not limited to Latin America. The 1972 Report of the British Branch of the International Law Association, written in April 1972, observes that the proposal "seems to contain the seeds of a possible compromise" in that it appeals to both developing States concerned about foreign fishing off their coasts and developed States' fears of "creeping jurisdiction." The statement further notes: "It is evident that the Venezuelan proposal, which makes a clear distinction between the territorial sea and areas beyond it in which other and lesser forms of jurisdiction are exercised, has attracted support for a distance criterion of 200 miles measured from the base lines from which the territorial sea is delimited, coupled with a depth criterion in respect of existing rights over continental shelves."[41] Subsequently, the base of international support for the Venezuelan proposal of 1971 has broadened substantially. The United States presented draft articles on a 200-nautical mile economic zone at the U.N. Conference on August 8, 1974.[42]

In June 1972, at the initiation of the Government of Colombia, a special conference of Caribbean countries was called to discuss various problems of the sea. Fifteen countries took part. On June 9, 1972, the Conference adopted the Declaration of Santo Domingo by 10 votes in favor, none against, and 5 abstentions.[43] The Declaration contained the texts of agreements reached on the territorial sea, the patrimonial sea, the continental shelf, the international seabed, the high seas, marine pollution, and regional cooperation. The patrimonial sea, according to the Declaration, is a zone adjacent to the territorial sea the breadth of which should be fixed by international agreement but is not to exceed a maximum of 200 nautical miles. In

---

40. U.N. Doc. A/AC. 138/SR. 64, at 3 (Aug. 12, 1971).
41. INTERNATIONAL LAW ASSOCIATION, REPORT OF THE BRITISH BRANCH COMMITTEE ON DEEP SEA MINING 14–16 (1972).
42. U.N. Doc. A/Cont. 62/C.2/L.47 (Aug. 8, 1974).
43. The countries in favor were Colombia, Costa Rica, Dominican Republic, Guatemala, Haiti, Honduras, Mexico, Nicaragua, Trinidad–Tobago, and Venezuela. The countries abstaining were Barbados, El Salvador, Guyana, Jamaica, and Panama.

that zone "[t]he coastal State has sovereign rights over the renewable and non-renewable natural resources . . . found in the waters, in the seabed and in the subsoil" thereof.[44]

On August 10, 1972, the United States representative to the Seabed Committee, Ambassador John Stevenson, made a statement indicating that the United States was prepared to acquiesce in the general idea of a resource zone in which the coastal State should have substantially complete jurisdiction with respect to seabed resources. He said:

[I]n order to achieve agreement, we are prepared to agree to broad coastal State economic jurisdiction in adjacent waters and seabed areas beyond the territorial sea as part of an overall law of the sea settlement. However, the jurisdiction of the coastal State to manage the resources in these areas must be tempered by international standards which will offer reasonable prospects that the interests of other States and the international community will be protected. It is essential that coastal State jurisdiction over fisheries and over the mineral resources of the continental margins be subject to international standards and compulsory settlement of disputes.[45]

The resource zone proposal also appears to be consistent with the 1972 Report of the Australian Branch of the International Law Association, which recommended "[o]n the question of the dividing line between the area subject to national jurisdiction and the international sea bed area," a " 'distance from the shore' criterion as part of the definition of the area within national jurisdiction."[46] The writers note that the 200-mile proposal, whether called "resource zone," "economic zone," or "patrimonial sea," does not require, with respect to seabed resources, that the geological formations of the submarine area within this zone be identical with, or even markedly similar to,

44. The text of the Declaration of Santo Domingo may be found in 11 INT'L LEGAL MATERIALS 892–93 (1972) *and as* U.N. Doc. A/AC.138/80 (July 26, 1972). *See also* COMM. ON THE PEACEFUL USES OF THE SEA-BED AND THE OCEAN FLOOR BEYOND THE LIMITS OF NATIONAL JURISDICTION, REPORT, 27 U.N. GAOR Supp. 21, at 70, 71–72, U.N. Doc. A/8721 (1972).

45. In August 1970, the United States had tabled a "working paper" in the form of a Draft Treaty that proposed that the coastal State, beyond the 200-meter line, should have substantial control over a zone from that isobath to the outer edge of the continental margin including the continental rise, denominated as a "trusteeship zone." U.N. Doc. A/AC. 138/25 (Aug. 3, 1970). In August 1972, Stevenson proposed five conditions, relating to noninterference with other uses of the oceans, international standards relating to pollution and protection of integrity of investments, revenue sharing, and compulsory settlement of disputes, but these conditions do not affect the jurisdictional problem discussed here. On June 20, 1974, Stevenson offered U.S. extension to a 12-mile territorial sea and U.S. agreement to a wider coastal State resource zone in exchange for international navigation guarantees, Wash. Post, Jun. 21, 1974, §A, at 3, col. 1.

46. ILA, *supra* note 26, at 2–3.

the geological formations of the land territories of the coastal State. The concept is one of a prolongation of jurisdiction, rather than one of jurisdiction related to a prolongation of land territories under water.[47] This is a significant distinction.

Conclusion as to the Limits of National Jurisdiction

The coastal States, in the writers' opinion, have exclusive competence, under customary international law as well as under the Convention, to govern the exploration and exploitation of the mineral resources of the seabed and subsoil of the submarine areas adjacent to their coasts that constitute the natural prolongation of their land territories, and this exclusive jurisdiction encompasses, and is limited by, the continental margin, irrespective of depth of water or distance from shore. The present writers believe that this jurisdiction under both conventional and customary law includes, *ab initio,* control of exploratory wells, so that no one may drill in the continental margin except under license of the appropriate coastal State.

It seems clear that the test of whether the continental margin is a prolongation of the adjacent land territory should relate to geomorphic, rather than geologic, criteria. It is unthinkable that jurisdiction of the adjacent coastal State over its continental slope should be displaced, retroactively, by drilling that may disclose that a particular formation so penetrated does not extend under the State's land territories, notwithstanding that the formation in question is part of the continental slope adjoining and buttressing those territories. This is not what the American States demanded at Ciudad Trujillo, and won at Geneva, when they insisted on their jurisdiction over the whole continental terrace, shelf and slope, down to "the greatest depths."

With respect to semienclosed seas, as a special case, it is suggested

47. For refinements of the patrimonial sea idea, see the Yaoundé seminar recommendations, U.N. Doc. A/AC.138/79 (Jul. 21, 1972), *reproduced in* COMM. ON THE PEACEFUL USES OF THE SEA-BED AND THE OCEAN FLOOR BEYOND THE LIMITS OF NATIONAL JURISDICTION, REPORT, 27 U.N. GAOR, Supp. 21, at 73, U.N. Doc. A/8721 (1972), 12 INT'L LEGAL MATERIALS 210, 212 (1973); *Draft articles on Exclusive Economic Concept, Submitted by Kenya,* U.N. Doc. A/AC. 138/SC. II/L. 10 (Aug. 7, 1972), *reproduced in* COMM. ON THE PEACEFUL USES OF THE SEA-BED AND THE OCEAN FLOOR BEYOND THE LIMITS OF NATIONAL JURISDICTION, REPORT, 27 U.N. GAOR, Supp. 21, at 180, U.N. Doc A/8721 (1972), 12 INT'L LEGAL MATERIALS 33 (1973); *Colombia, Mexico and Venezuela: draft articles of treaty,* U.N. Doc. A/AC.138/SC.II/L.21, at 2 (Apr. 2, 1973), 12 INT'L LEGAL MATERIALS 570 (1973); O.A.S. Inter-American Juridical Comm. Resolution on The Law of the Sea, 12 INT'L LEGAL MATERIALS 711 (1973).

that the prolongation criterion of the *North Sea Continental Shelf Cases* may be satisfied adequately if, first, the surface relationship of an island arc to a continent is such as to characterize the sea thus enclosed as constituting a geographic unit, and, second, the island arc is separated from the abyssal ocean floor by definite geomorphic indices, such as a long and abyssal trench, the Marianas Trench or the Ryukyu Trench for instance, as distinct from a local trough, such as the Norwegian Trough. In such case, a sea thus enclosed by an island arc is to be deemed, for purposes of jurisdiction with respect to seabed resources, as constituting an area adjacent to, and a prolongation of, all of the mainland and island territories which are littoral to that semienclosed sea. This extrapolation of the prolongation doctrine to the semienclosed sea may be in some cases a substantial one spatially. But in our view there is no logical alternative to it, once the prolongation doctrine is held applicable to the whole continental margin, as the writers believe it must be.[48]

Theory and Practice Applied to the South China Sea

The South China Sea: Jurisdictional Testing Ground

The South China Sea provides a rigorous test of the extent of national seabed jurisdiction in the marginal sea, thereby shedding light on the general problem of the limits of national seabed jurisdiction worldwide.

Geographically, the South China Sea is a large body of water on the southeastern shore of the Asian continent. It is set off from the Pacific by Taiwan, the Philippine Islands, Brunei, certain of the Indonesian islands, and those of Malaysia. It is one of a number of semienclosed seas that skirt the whole eastern edge and southeastern corner of Asia, flanked by a chain of island arcs. From north to south, these Asian semienclosed seas include the Bering Sea, the Sea of Okhotsk, the Sea of Japan, the East China Sea, the South China Sea, the Andaman Sea, and the various Indonesian archipelagic seas. The outer island arcs of these semienclosed seas border them on their easternmost, or Pacific, sides. Recent scientific research[49] indicates

---

48. *See* GULF AND CARIBBEAN MARITIME PROBLEMS (L. Alexander, ed. 1972) (Law of the Sea Workshop, Law of the Sea Institute, University of Rhode Island).

49. Matthews, *This Changing Earth,* 143 NAT'L GEOGRAPHIC 1 (1973); Dietz & Holden, *Reconstruction of Pangaea: Breakup and Dispersion of Continents, Permian to Present,* 75 J. GEOPHYSICAL RESEARCH 4939 (1970); Bullard, *The Origin of the Oceans,* 221 SCIENTIFIC AM. 66 (1969); Emery, *Relict Sediments on*

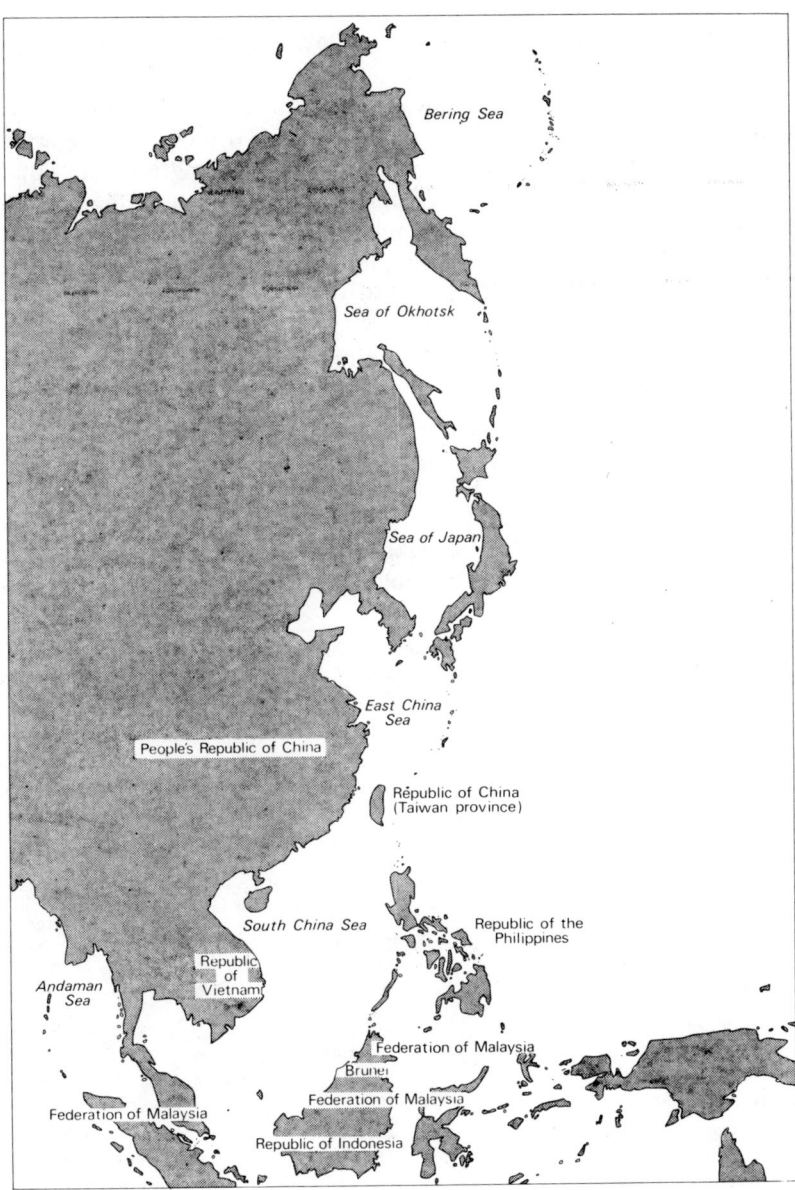

5. Asian semienclosed seas

that these island arcs are the submerged geological outposts of the Asian continental landmass. The writers may anticipate the argument somewhat by describing the semienclosed South China Sea as a sea overlying the Asian continental margin, or, more easily, as a marginal sea.

The width of the South China Sea, measured from the Vietnamese coast near Phanrang through the Dangerous Ground to Balabac Island, south of Palawan, is nearly 600 miles. The length of the South China Sea from the Republic of China (Taiwan province) southwest to the Singapore Strait is approximately 1,800 miles. Including the Gulfs of Thailand and Tonkin for geographical measurement only, the area of the South China Sea is 1,148,500 square miles, with an average depth of 4,802 feet (1,464 meters).[50] Of the world's fourteen major seas, the South China Sea is the largest in area and the sixth deepest in average depth.[51]

Seven littoral States are involved. Clockwise from the Formosa Strait are the Republic of China (Taiwan province), the Republic of the Philippines, the Federation of Malaysia, Brunei, Malaysia again, the Republic of Indonesia, Malaysia's western coast, the Republic of Vietnam, and the People's Republic of China. Arbitrarily excluded from this discussion are the Republic of Singapore, because of its re-

---

Continental Shelves of World, 52 AM. ASS'N PETROLEUM GEOLOGISTS BULL. 455 (1968); F. SHEPARD, SUBMARINE GEOLOGY (1963); SECRETARIAT, UNITED NATIONS EDUCATION, SCIENTIFIC & CULTURAL ORGANIZATION, *Scientific Considerations Relating to the Continental Shelf,* U.N. Doc. A/Conf. 13/2 and Add. 1 (Sept. 20, 1957) (Preparatory Document No. 2); Hess, *Major Structural Features of the Western North Pacific, an Interpretation of H.O. 5485, Bathymetric Chart, Korea to New Guinea,* 59 GEOLOGICAL SOC'Y AMERICA BULL. 417 (1948).

50. NATIONAL GEOGRAPHIC ATLAS OF THE WORLD 7 (1970).

51. *Id.* The fourteen major seas are:

Sea	Area in square miles	Average depth in feet & meters
South China Sea	1,148,500	4,802 (1,464m)
Caribbean Sea	971,400	8,448 (2,575m)
Mediterranean Sea	969,100	4,926 (1,501m)
Bering Sea	873,000	4,893 (1,491m)
Gulf of Mexico	582,100	5,297 (1,615m)
Sea of Okhotsk	537,500	3,192 (973m)
Sea of Japan	391,100	5,468 (1,667m)
Hudson Bay	281,900	305 (93m)
Andaman Sea	218,100	3,667 (1,118m)
Black Sea	196,100	3,906 (1,191m)
Red Sea	174,900	1,764 (538m)
North Sea	164,900	308 (94m)
Baltic Sea	147,500	180 (55m)
Caspian Sea	143,550	3,264 (995m)

cessed location in the Singapore Strait; Thailand and the Khmer Republic (Cambodia), because they do not extend past the mouth of the Gulf of Thailand; and the Democratic Republic of Vietnam (North Vietnam), because we regard it as bordering solely on the Gulf of Tonkin. The South China Sea, for the purpose of this discussion, is defined as that body of water south of the Formosa Strait, west of the Luzon, Mindoro, and Balabac Straits, north of the equator and the Singapore Strait, and east of the Gulfs of Thailand and Tonkin.

There are many hundreds of islands, islets, and rocks in the South China Sea. Many of these islands are claimed by more than one State. Historically, these claims are complex, contradictory, and difficult to document. Claims were made by European imperial powers as well as the South China Sea littoral States.[52] The two principal island groupings are, first, the Paracel Islands–Macclesfield Bank group in the northwestern quadrant of the South China Sea, in the vicinity of 15° to 17° north latitude and 111° to 115° east longitude; and, second, the larger area of the Dangerous Ground, including the Spratly Islands. The Dangerous Ground long has been a hazard to shipping. Until recent years it was one of the world's least accurately charted marine areas. Specifically, the Dangerous Ground lies in the southeastern quarter of the South China Sea, between 7° and 12° north latitude and 112° and 118° east longitude, and generally between the Vietnamese Mekong delta and Palawan Island of the southern Philippines.

No question of national seabed boundary delimitation yet has arisen that approaches in complexity the legal, political, and geological situa-

52. There appears to be no single unclassified source of information on claims to South China Sea islands. Historically, Imperial China claimed the whole South China Sea, and the People's Republic of China may do so still. The British at one time claimed portions of the Spratly Islands but are understood to have abandoned their claims. When France possessed Indochina, it made a similar claim, Notice of July 26, 1933, [1933] J.O. 7837, and is understood to have relinquished it when it forsook sovereignty over the Vietnamese mainland. These impressions are partly distilled from the files of the Office of the Geographer, U.S. Dep't of State. The best publication on South China Sea islands, if somewhat slanted, is *The Vietnamese Islands of Paracels and Spratly,* 5 PRESS & INFORMATION OFFICE, REPUBLIC OF VIETNAM (Mar. 16, 1959) (No. 6). In 1974 the South China Sea islands dispute erupted. In January 1974 the Republic of Vietnam lost a two-day battle with the People's Republic of China for the Paracel Islands; the Vietnamese then promptly sent 200 troops to defend the Spratly Islands. Wash. Post, Feb. 1, 1974, §A, at 26, col. 3; THE ECONOMIST, Jan. 26, 1974, at 44, col. 1. Since the authors conclude *infra* that South China Sea seabed boundaries are to be determined without reference to midsea islands and islets, they disregard this farrago of claims. Speculation as to jurisdiction over East Asian islands is not confined to the South China Sea. *See* Cheng, *The Sino-Japanese Dispute Over the Tiao-Yu-Tai (Senkaku) Islands and the Law of Territorial Acquisition,* 14 VA. J. INT'L L. 221, 222 n.2 (1974).

tion in the South China Sea. The *North Sea Continental Shelf Cases* involved three Western European States, two of which had ratified the 1958 Geneva Continental Shelf Convention, and a body of water less than 200 meters deep, excepting the Norwegian Trough, which was not involved in the controversy.[53] But the South China Sea is bordered by seven States of varying size and cultural affinity, only two of which are parties to the Convention,[54] and reaches 5,000 meters maximum depth in the Manila Trench just west of Luzon.[55]

Do the coastal States littoral to the South China Sea, collectively, have competence, to the exclusion of any outside State and of a putative international regime, to control the exploration and exploitation of the resources of their seabeds, specifically, petroleum? If so, how are South China Sea seabed boundaries to be drawn? What effect shall be given to the hundreds of islands in midsea, many of which are involved in tortuous disputes over ownership?

The Geomorphology of the Seabed

The geomorphology of the South China Sea's floor should be visualized in four distinct provinces.[56] First, the Gulfs of Thailand and Tonkin, not considered here, are flat, shallow areas of less than 100 meters depth. Second, the flat, petroleum- and gas-rich Sunda Shelf under the southwest third of the South China Sea is less than 200 meters deep. Substantial petroleum production is under way in this area. Third, the Dangerous Ground of the China Basin, riddled with towering undersea volcanic intrusions, is about 2,000 meters deep. Fourth, the deep plain of the China Basin exceeds 4,000 meters depth in some places. Other than these four provinces, major features of the South China Sea's submerged topography include the Palawan Trough, almost 3,000 meters deep and found between the Balabac Strait, south of the Republic of the Philippines, and the Dangerous Ground; the West Luzon Trough, 2,600 meters deep; and the Manila Trench, 5,000 meters deep, lying west of Luzon. In summary, the

---

53. Denmark became party to the Continental Shelf Convention on June 12, 1963, and the Netherlands on Feb. 18, 1966. Office of the Geographer, Bureau of Intelligence & Research, U.S. Dep't of State, Limits in the Seas, National Claims to Maritime Jurisdictions, No. 36, at 25, 81 (1974).
54. The Republic of China (Taiwan province) became party to the Continental Shelf Convention, with important reservations, on Oct. 12, 1970, and the Federation of Malaysia on Dec. 21, 1960. *Id.* at 97, 72.
55. Emery, *supra* note 3, at 843.
56. *See generally* Emery, *supra* note 3.

three geomorphological provinces of the South China Sea floor that concern us here, indicated on Map 6 in pocket part (see also Appendix 1), are the Sunda Shelf, the Dangerous Ground, and the deep plain.

South China Sea littoral States differ in the rate of steepness with which their submerged continental shelves and slopes descend to greater depths.[57] The eastern Federation of Malaysia south of Brunei, the Republic of Indonesia, western Malaysia, the Republic of Vietnam, and the People's Republic of China generally enjoy shallow depths for a considerable distance offshore. The major manifestation of this topographic fact is the Sunda Shelf. On the other hand, the Republic of China (Taiwan province), the Republic of the Philippines, eastern Malaysia north of Brunei, and Brunei itself possess continental shelves and slopes that drop off suddenly to greater depths.

The Sunda Shelf is the relatively shallow extension of the southeast Asian continental landmass, connecting the Federation of Malaysia and the Republic of Indonesia with the Republic of Vietnam. Bathymetric mapping offshore of southeast Asia demonstrates visually the structural continuity evidenced by geological data. The Sunda Shelf is the Asian continent, only under a few meters of water. According to Emery and Ben-Avraham, "[t]he basement rocks of the Indochina Peninsula, the Malay Peninsula, Sumatra, and Borneo continue beneath the shallow Sunda Sea, and they consist mainly of Paleozoic and Mesozoic metamorphic and igneous rocks."[58] In claiming seabed natural resources jurisdiction over the Sunda Shelf, the States littoral to this southwestern third of the South China Sea are clearly asserting jurisdiction over their own national submerged continental landmasses.

57. *Shelf* and *slope*, as well as *margin* and *rise*, are terms with specific geomorphological meanings in describing the submerged areas off the continental shoreline. A clear explanation of these terms is found in REPORT ON MARINE SCIENCES AND TECHNOLOGY, CMND. NO. 3992, at 51 (1969) (presented to the United Kingdom Parliament): "The Shelf is normally inclined at an angle of about 1/10°. Where there is an increase in gradient above 1 1/2°, the Slope is said to have begun. Where it subsequently falls below 1 1/2°, from the average of between 3° and 6°, the Rise is said to have begun. The Rise continues down to the deep ocean floor or Abyssal Plain. The widths and depths of all three features are very variable. Generally the Shelf extends to a depth between 130 and 200 metres, but, exceptionally, to between 50 and 500 metres. The foot on the Slope may vary from 1,500 to 4,000 meters in depth and the Rise generally continues down to between 4,000 and 5,000 metres. *The Continental Shelf, Slope and Rise are collectively known as the Continental Margin.*" (emphasis supplied). *See also* Emery, *Characteristics of Continental Shelves and Slopes,* 49 AM. ASS'N PETROLEUM GEOLOGISTS BULL. 1379 (1965).

58. Emery, *supra* note 3, at 845. See also Ben-Avraham & Emery, *Structural Framework of Sunda Shelf,* 57 AM. ASS'N PETROLEUM GEOLOGISTS BULL. 2323 (1973).

Recent scientific investigations have supported the theory that the Dangerous Ground is a single geological plate intruded by numerous volcanic formations in more recent geologic time. This platform has been divided from the Palawan and Borneo formations by the Palawan Trough. Its volcanic intrusions rise like skyscrapers from the sea floor to break the water surface as islands, islets, and rocks. A recent description of the Dangerous Ground reveals its complexity:

The bottom descends from the northern Sunda Shelf and its extensions via a continental slope that contains a broad terrace near the Indochina Peninsula. Sediments from the northern Sunda Shelf evidently have been carried down the southwestern part of the continental slope and have built small continental rises or aprons that have partly buried irregular topography between provinces that have been termed benches. The largest remaining parts of these benches are off the narrow shelf east of the Indochina Peninsula. Off northern Borneo the base of the continental slope is bordered by a narrow deep area (maximum depth 2,944 m), the Palawan trough. Beyond the trough and intruded by sediments from the aprons south of the trough is a broad, very irregular plateau ("hills" . . .). Some of the numerous seamounts within this region rise above the surface as islands and rocks, many of which are fringed by reefs; the descriptive phrase "Dangerous Ground" has long been applied to this region.[59]

The Dangerous Ground is surrounded by the deep sedimentary deposits of the Asian continental slope on the west, the Sunda Shelf on the south, and the Palawan Trough on the east.

Scientific research indicates that the Dangerous Ground is composed of the same rocks as the continental Sunda Shelf, giving it also a basically continental, not oceanic, physical character. Emery and Ben-Avraham observe that the local volcanic peaks, huge coral reefs,

---

59. Parke, *et al., Structural Framework of Continental Margin in South China Sea,* 55 AM. ASS'N PETROLEUM GEOLOGISTS BULL. 723, 726 (1971), *also found at* U.N. ECON. COMM'N for ASIA & THE FAR EAST, COMM. FOR COORDINATION OF JOINT PROSPECTING FOR MINERAL RESOURCES IN ASIAN OFFSHORE AREAS, 4 TECHNICAL BULL. 103 (1971). Valuable recent research into South China Sea geology includes Kim, *Prospective Oil Fields on the Continental Shelf in Eastern Asia and Some Associated Political Problems,* 3 PROC. ASS'N AM. GEOGRAPHERS 93 (1971); Wageman, *et al., Structural Framework of East China Sea and Yellow Sea,* 54 AM. ASS'N PETROLEUM GEOLOGISTS BULL. 1611 (1970); Hedberg, *Continental Margins from Viewpoint of the Petroleum Geologist,* 54 AM. ASS'N PETROLEUM GEOLOGISTS BULL. (1970); Emery, *et al., Geological Structure and Some Water Characteristics of East China Sea and Yellow Sea,* U.N. ECON. COMM'N FOR ASIA & THE FAR EAST, COMM. FOR COORDINATION OF JOINT PROSPECTING FOR MINERAL RESOURCES IN ASIAN OFFSHORE AREAS, 2 TECHNICAL BULL. 3 (Japan: Geological Survey of Japan, 1969); Niino & Emery, *Sediments of Shallow Portions of East China Sea and South China Sea,* 72 GEOLOGICAL SOC'Y AM. BULL. 731 (1961).

and older folded sediments are probably the northern "continuation of pre-Cenozoic igneous and metamorphic rocks beneath the Sunda Shelf."[60] The area's irregular geology, according to them, also implies the presence of petroleum deposits: "Ridges surround the basin, where they served as submerged dams to trap large quantities of detrital sediments brought to the ocean by rivers. One of the ridges bordering the shelf off Borneo continues northeastward as the elongate Palawan Island. These barriers appear to be close parallels with the ones previously observed in the East China Sea. . . ."[61] Thus, the geology of the Dangerous Ground indicates a sedimented, folded, volcanized area of continental rock under a deep water column. If geology were to control, States littoral to this third of the bed of the South China Sea might assert legal rights to the natural resources of this predominantly continental seabed area.

The deep plain of the China Basin is found in the northeastern third of the South China Sea. The deep plain is flat, with a greatest depth of 4,350 meters.[62] The depth and flatness of this plain indicate a more oceanic, less continental quality. Nevertheless, compared to the truly oceanic depths of the Philippines' Pacific coast, over 10,000 meters, the depth of this plain becomes less impressive. Also, the deep plain appears to be separated from truly oceanic areas by a formidable ridge. Emery and Ben-Avraham state: "The Central Range of Taiwan . . . appears to be a northerly continuation of a ridge that, farther south, separates the Manila Trench and the West Luzon Trough, and may continue southeastward through the central Philippine island of Mindoro. . . ."[63] It has been suggested that these ridges are being pushed up by the action of the immense, oceanic Philippines Basin Plate underthrusting the continental structures of the South China Sea floor to its west.

The Seabed of the South China Sea: as the Prolongation of the Landmass of the Asian Continent

In sum, scientific writings lean toward the conclusion that the South China Sea overlies the continental margin of southeastern Asia,[64]

---

60. Emery, *supra* note 3, at 843.
61. *Id.* at 845.
62. *Id.* at 841.
63. *Id.* at 856.
64. Emery, *et al., Geological Structure and Some Water Characteristics of the Java Sea and Adjacent Continental Shelf,* U.N. ECON. COMM'N FOR ASIA & THE FAR EAST, COMM. FOR COORDINATION OF JOINT PROSPECTING FOR MINERAL

treating these areas on a geomorphic basis. Indeed, if the problem is considered solely as one of geology, scientists today postulate a continental nature for much of the bed of the South China Sea, but not all. This is not unusual. The margins of the continents are poorly explored. As Burk observes, they are the "transition between oceanic and continental crust."[65] The new global tectonics of continental drift, sea-floor spreading, and mantle convection suggest that the geological nature of the seabed of the South China Sea is not a simple matter of surface distance from the shoreline or of depth measurement. Two-thirds of the South China Sea floor is continental, and that which is not continental is not truly oceanic.

If, as seems to be the better scientific argument, the boundary of the Asian landmass is to be deemed the line of deep trenches seaward of the island arcs flanking nearly the whole coast of the continent, then the seabed enclosed between the continent and its arcs of islands must be deemed the prolongation of these exposed continental features. Or, to put it the other way around, the islands lying between the coast of the continent and the deep trenches, such as the Ryukyu and Mariana Trenches located seaward of those island arcs, are but the surface projections of the continental landmass that extends out to those deep oceanic trenches. On this hypothesis, the seabed of the South China Sea, being the submerged component of this continental complex, is the prolongation of the continent just as are the exposed portions: the island arcs. This being so, the applicability of the prolongation doctrine of the *North Sea Continental Shelf Cases* would seem persuasive. That doctrine, while articulated in a case that involved waters less than 200 meters deep, is not limited by depth or distance. Indeed, as to distance, the North Sea is some 400 miles wide in places, and the South China Sea's greatest width exceeds this by only about 50 percent. As to depth, the rationale of the Court's opinion in the North Sea case would have been equally compelling if the North Sea had happened to be several times as deep as it is.

Adjacency Criterion and the South China Sea

The exploitability question on a superficial level seems to answer itself. For example, there is no practical problem, hence no real legal

RESOURCES IN ASIAN OFFSHORE AREAS, 6 TECHNICAL BULL. 197 (1972); Karig, *Origin and Development of Marginal Basins in the Western Pacific*, 76 J. GEOPHYSICAL RESEARCH 2542 (1971); Menard & Smith, *Hypsometry of Ocean Basin Provinces*, 71 J. GEOPHYSICAL RESEARCH 4305 (1966).
65. *See* note 37 *supra*.

one, if the areas in question are not exploitable because technology at a given time does not make possible the drilling of wells. On the other hand, if technology has advanced to the point of making oil well drilling possible, the area, by hypothesis, is one where, in the Convention's language, the depth of the superjacent waters admits of exploitation of the seabed and subsoil. Depth, therefore, is not a disqualifying factor.

The true question, which is a difficult one, is whether all areas of the seabed of the South China Sea can be said to be adjacent to some coastal State. It must always be borne in mind that the mere fact of the exploitability of an area does not confer seabed jurisdiction on a coastal State, unless that area is adjacent to that State. Earlier in this chapter adjacency was equated with the International Court of Justice's concept of prolongation of land territories, rejecting, as did the Court, the equation of *adjacency* with *proximity*. In the *North Sea Continental Shelf Cases,* the parties, pursuant to the Court's Opinion, deemed an area some 194 miles from Germany's coast as being available for the exercise of Germany's seabed jurisdiction under principles of customary law. The median line in the South China Sea, in the extreme instance, is approximately 300 miles from the coast of a major body of land. The difference between these numbers is not overwhelming.

Application of the *Semienclosed Sea* Variation

The geographical enclosure of any sea is a relative matter. The term *semienclosed sea* is used here to describe geographically those bodies of water encircled by islands and covering substantial portions of the world's continental margins. By geologic and geomorphic tests, the South China Sea is a marginal sea, while geographically, it is a semienclosed sea. As previously noted, along the eastern coast of Asia these semienclosed seas are bounded on one side by the continental shoreline and on the other by the elongated system of island arcs, actually the surfaced tops of long ridges, of the Asian continental margin. The States littoral to the South China Sea form a continuous annular chain around that body of water.

The primary argument against recognizing the exclusive character of the interests of the littoral States, as a whole, is the size of the sea they enclose. The same argument, which is one of degree, might be made as to the Gulf of Mexico or the Caribbean Sea. But, viewed on a global scale, the South China Sea, the Caribbean Sea, and the Gulf of Mexico, like all other marginal seas of the world, are distinguished by relatively narrow straits that give access to them from the oceans. Put another way, they are distinguished by the contiguity of peripheral landmasses to one another. The lands around the South China Sea bear a peculiar importance to one another, and hence to the sea that they share, whether the relationship is measured in terms of national security, trade, or fishing, or whether they are considered as land bases for seaborne exploitation of submarine minerals. In questioning the extent of the exclusive seabed jurisdiction of South China Sea littoral States, one would be rash to ignore the special applicability of the criteria stated in the 1945 Truman Proclamation to the whole area.

Application of the Concept of the Patrimonial Sea

The point need not be labored. A glance at Map 6 in pocket part will show the relatively small area in midsea that would be exterior to, and surrounded by, a 200-mile resource zone, measured from the peripheral coasts. Internationalization of the small midsea remnant would be impractical. Since the South China Sea is the largest marginal sea in the world, we conclude that, a fortiori, the States littoral to the remainder of the world's marginal seas should share seabed resource jurisdiction to the exclusion of any international interest.

Conclusion: Competence of South China Sea States to Control Exploration and Exploitation

That the seabeds of the Asian marginal seas enclosed by island arcs must be deemed, along with their enclosing islands, the prolongation of the Asian continental landmass seems a reasonable scientific conclusion. On this hypothesis, the rationale of the *North Sea Continental Shelf Cases* would strongly support, if not compel, the legal conclusion that the States occupying the Asian island arcs must be accorded competence, correlative with that of mainland States, to govern the exploration and exploitation of the seabed resources of the marginal seas they enclose. The authors believe that public international law will evolve in this direction, unless present customary international law and the premises of the *North Sea Continental Shelf Cases* are overturned by a new Law of the Sea Convention in the near future. As to that event, the trend of recent debates in the United Nations Seabed Committee as well as the opening plenary session speeches of many States favoring a 200-mile zone at the Third United Nations Conference on the Law of the Sea, Caracas, Venezuela, in

June and July 1974, makes it unlikely that coastal States will abandon their national interest in extensive seabed jurisdiction.[66]

Principles of Seabed Boundary Demarcation

Question of Common Prolongation

As the Court in the *North Sea Continental Shelf Cases* made clear, the law does not concern itself with determining the "fair share" of the seabed that should be allocated to a coastal State, as though appropriating territory not previously owned by anyone. To the contrary, the problem of demarcation of seabed boundaries between coastal States does not arise at all unless it is first determined that the area sought to be delimited constitutes the common prolongation of the land territories of both of them.[67] This is a sound protection against an exuberant effort by coastal States on opposite sides of an ocean to carve up the whole seabed. Because of this underlying premise or condition, the present writers have examined above, at some length, the troublesome question of whether, indeed, the seabed of the whole South China Sea properly can be considered to appertain to the littoral States, and to them alone. The writers have answered this question affirmatively. The problem now arises of determining seabed boundaries among those coastal States, assuming the absence of agreement and hence the necessity for a third-party delimitation in arbitration or litigation.

The Equidistance Principle

Seabed boundary problems are of two kinds: those relating to frontal boundaries between States occupying opposite coasts and those relating to lateral boundaries between States occupying adjoining portions of the same coast. The South China Sea presents both situations.

Boundaries, of course, may be set by agreement in any manner the parties select, and there are examples of such agreements in the South

---

66. *See* the Yaoundé seminar recommendations and the Kenyan Draft Articles, on the theme of the patrimonial sea, note 47 *supra*. *See also* Statement of Chairman, People's Republic of China delegation, U.N. Conference (ECAFE), Colombo, Sri Lanka (Apr. 2, 1974, translation): "Division of jurisdiction of the continental shelf between China and countries bordering on or facing her should be decided by the countries concerned through consultations on an equal footing."

67. [1969] I.C.J. 3, 31 (para. 43).

China Sea.[68] But in the absence of agreement, what criteria should an international tribunal apply; and, if an agreement is to be negotiated, do the precedents of other agreements suggest any generally applicable criteria? A number of guidelines have been used or suggested in the demarcation of seabed boundaries. Of these, the equidistance principle has been given the widest recognition, whether by its direct application or by acknowledgment that only the existence of special circumstances justifies an exception to the rule.

The equidistance-line technique, and the reasons for it, have been summarized as follows by the Geographer of the United States State Department:

A median line (at times called "lateral line") has proved to be the best solution for delineating water areas between sovereignties. In both theory and practice the geometrical principle involved in determining the median line is the most satisfactory which has so far been devised, lending itself admirably to the construction of equitable boundaries between states. It depends upon precise measurement rather than subjective factors. Without delving into its technical characteristics, a median line is defined as a line, or boundary, every point of which is equidistant from the nearest points on the lines from which it is measured. Oddly enough, the technique upon which the construction of such lines depends is purely trial and error, that is, establishment of points contingent upon being so placed that they be no farther from one than from the other fixed point representing the two sovereignties.

. . . .

The spirit of the articles on median lines is to provide a means whereby boundary agreements between states may be facilitated. *But since median-line boundaries are objective they can frequently be used at least as a point of departure in the reaching of agreement.* Site of known or potential resources, location of a navigation channel, or traditional offshore practices of a state are among special circumstances which may give rise to modifying or even disregarding completely a median line in affixing a boundary. For example, a boundary in the territorial sea may only roughly approximate a median line, compensating for loss of an area in one place by gain in another. Despite such departures from a formula the actual precisely constructed median line stands as a potential means of establishing fair and lasting offshore boundaries.[69]

68. The legislation and decrees of the Republic of the Philippines, the Federation of Malaysia, and the Republic of Indonesia recognize the necessity of determining seabed boundaries according to equitable principles. *See, e.g.,* U.N. Doc A/ AC.135/11, at 47 (1968). *See also* Agreement Between Indonesia and Malaysia Relating to the Delimitation of the Continental Shelves Between the Two Countries, Oct. 27, 1969, 9 INT'L LEGAL MATERIALS 1173 (1970).

69. Office of the Geographer, Bureau of Intelligence & Research, U.S. Dep't of State, Geographic Bull. No. 3, Sovereignty of the Sea, App. B, 15–16 (rev. Oct. 1969) (emphasis supplied). For a more detailed description of the geometrical techniques for drawing median lines, see I SHALOWITZ, SHORE AND SEA BOUNDARIES 230–35, 253–54 (1962) (U.S. Dep't of Commerce).

The median line, or equidistance principle, is endorsed in Article 6 of the Convention on the Continental Shelf, in the absence of agreement or special circumstances. It deals separately with the problems of seabed boundaries between opposite coasts and those between adjacent States on the same coast. It provides:

1. Where the same continental shelf is adjacent to the territories of two or more States whose coasts are opposite each other, the boundary of the continental shelf appertaining to such States shall be determined by agreement between them. In the absence of agreement, and unless another boundary line is justified by special circumstances, the boundary is the median line, every point of which is equidistant from the nearest points of the baselines from which the breadth of the territorial sea of each State is measured.

2. Where the same continental shelf is adjacent to the territories of two adjacent States, the boundary of the continental shelf shall be determined by agreement between them. In the absence of agreement, and unless another boundary line is justified by special circumstances, the boundary shall be determined by application of the principle of equidistance from the nearest points of the baselines from which the breadth of the territorial sea of each State is measured. . . .

In the *North Sea Continental Shelf Cases* the International Court of Justice had before it the question of lateral boundaries between Germany and Denmark and between Germany and the Netherlands. One issue was whether the second paragraph of Article 6 of the Convention, which deals with lateral boundaries, stated customary law with regard to the boundaries between States adjacent to one another on the same coast. The question was important because Germany had not ratified the Convention. The Court's Opinion drew a sharp distinction between seabed boundaries between States occupying opposite coasts and those between adjacent States on the same coast. As to the former, the Opinion implied that the median line principle is to be recognized as one of customary international law. As to the latter, it reached the opposite conclusion. The reasons given for the distinction were these:

Before going further it will be convenient to deal briefly with two subsidiary matters. Most of the difficulties felt in the International Law Commission related, as here, to the case of the lateral boundary between adjacent States. Less difficulty was felt over that of the median line boundary between opposite States, although it too is an equidistance line. For this there seems to the Court to be good reason. The continental shelf area off, and dividing, opposite States, can be claimed by each of them to be a natural prolongation of its territory. These prolongations meet and overlap, and can therefore only be delimited by means of a median line; and, ignoring the presence of islets, rocks and minor coastal projections, the disproportionally distorting effect of which can be eliminated by other means, such a line must effect an equal division of the particular area involved. If there is a third State on one of the coasts concerned, the area of mutual natural prolongation with that of the same or another opposite State will be a separate and distinct one, to be treated in the same way. This type of case is therefore different from that of laterally adjacent States on the same coast with no immediately opposite coast in front of it, and does not give rise to the same kind of problem—a conclusion which also finds some confirmation in the difference of language to be observed in the two paragraphs of Article 6 of the Geneva Convention (reproduced in paragraph 26 above) as respects recourse in the one case to median lines and in the other to lateral equidistance lines, in the event of absence of agreement.

If on the other hand, contrary to the view expressed in the preceding paragraph, it were correct to say that there is no essential difference in the process of delimiting the continental shelf areas between opposite States and that of delimitations between adjacent States, then the results ought in principle to be the same or at least comparable. But in fact, whereas a median line divides equally between the two opposite countries areas that can be regarded as being the natural prolongation of the territory of each of them, a lateral equidistance line often leaves to one of the States concerned areas that are a natural prolongation of the territory of the other.[70]

The care with which the Court's Opinion distinguished between the application of the median line as between opposite countries, to which it apparently gave implied approval as a feature of customary international law, and the principle of equidistance as between adjacent States, which it declined to approve as customary law,[71] leads to the conclusion that the Court would probably apply the median line concept as the primary point of customary law in a case involving the seabed boundary between opposite coasts and that to overcome this presumption strong proof of special circumstances would be required.

The practice of States has been overwhelmingly in support of the median line concept as between opposite coasts, in the demarcation of boundaries both in the territorial sea and in the seabed seaward of the territorial sea, under either conventional or customary law. The variations have been due to special circumstances, such as the presence of an islet owned by one nation off the coast of another.

Special Circumstances

Article 6 of the Convention on the Continental Shelf treats the equidistance principle as the rule; "special circumstances" as the ex-

70. [1969] I.C.J. 3, 36–37 (paras. 57–58).
71. With respect to the seabed boundary between Germany, the Netherlands, and Denmark, the Court was confronted by the controlling factor of the concave coastline of Germany. The consequence was that an equidistance line between that coast and the adjoining coast of Denmark, on the one hand, and a similar line between

ceptions. The Opinion in the *North Sea Continental Shelf Cases*, applying customary law, treats the equidistance concept as merely one of a number of equitable criteria, not the governing one, with respect to lateral boundaries. The Court's concise decree, paragraph 101, merits quotation in full:

> For these reasons,
> THE COURT,
> by eleven votes to six,
> finds that, in each case,
>
> (A) the use of the equidistance method of delimitation not being obligatory as between the Parties; and
>
> (B) there being no other single method of delimitation the use of which is in all circumstances obligatory;
>
> (C) the principles and rules of international law applicable to the delimitation as between the Parties of the areas of the continental shelf in the North Sea which appertain to each of them beyond the partial boundary determined by the agreements of 1 December 1964 and 9 June 1965, respectively, are as follows:
>
> (1) delimitation is to be effected by agreement in accordance with equitable principles, and taking account of all the relevant circumstances, in such a way as to leave as much as possible to each Party all those parts of the continental shelf that constitute a natural prolongation of its land territory into and under the sea, without encroachment on the natural prolongation of the land territory of the other;
>
> (2) if, in the application of the preceding sub-paragraph, the delimitation leaves to the Parties areas that overlap, these are to be divided between them in agreed proportions or, failing agreement, equally, unless they decide on a régime of joint jurisdiction, user, or exploitation for the zones of overlap or any part of them;
>
> (D) in the course of the negotiations, the factors to be taken into account are to include:
>
> (1) the general configuration of the coasts of the Parties, as well as the presence of any special or unusual features;
>
> (2) so far as known or readily ascertainable, the physical and geological structure, and natural resources, of the continental shelf areas involved;
>
> (3) the element of a reasonable degree of proportionality, which a delimitation carried out in accordance with equitable princi-

---

Germany and the Netherlands, on the other, would form a triangle that would pinch off Germany's share of the continental shelf far short of the median line between Germany and the United Kingdom, abandoning to Denmark and the Netherlands a considerable area of the seabed that in fact lies off the coast of Germany.

ples ought to bring about between the extent of the continental shelf areas appertaining to the coastal State and the length of its coast measured in the general direction of the coastline, account being taken for this purpose of the effects, actual or prospective, of any other continental shelf delimitations between adjacent States in the same region.[72]

The decision of the Court, in general, was thus generally favorable to Germany.[73]

To this list of what might be called special circumstances should be added the Court's earlier reference, in paragraph 57, to "islets, rocks and minor coastal projections, the disproportionally distorting effect of which [on the median line] can be eliminated by other means." Small islands constitute the most troublesome of all special circumstances in the calculation of median, or equidistance, lines. The problem has been stated clearly by Miss J. A. C. Gutteridge, at the time of the United Kingdom Foreign Office:

There is also some obscurity in Article 6 of the Convention as to what is meant by "special circumstances." One clear example of "special circumstances" is, however, the presence of islands. Where the continental shelf underlies an area of shallow sea, such as the Persian Gulf, which has many islands and is surrounded by the coasts of opposite or adjacent States, the drawing of the boundary on the strict principle of the median line could, it is clear, result in many curious and inequitable deflections of the median line. There may, for instance, be a very small island which lies approximately in the middle of the shallow sea; or there may be islands which are so close to the mainland as to be justifiably considered part of the mainland for the purposes of working out the boundary of the continental shelf. Again there may be islands which although near the coast of State A are under the sovereignty of State B. All these circumstances not only show the difficulty of a uniform application of the median line principle, but also explain why the 1958 Geneva Conference found itself unable (as did the International Law Commission in its draft Articles) to include in the Convention any specific provisions about the effect of the presence of islands on the delimitation of the boundaries of the continental shelf.[74]

---

72. [1969] I.C.J.3,53–54 (para. 101).
73. An agreement among Germany, the Netherlands, and Denmark subsequently was reached in accordance with the Court's decree. The effect was to allocate to Germany an area that, at its farthest extent, is some 194 miles from the German coast. 10 INT'L LEGAL MATERIALS 600–12 (1971) (*see* map at 602).
74. Gutteridge, *The 1958 Geneva Convention on the Continental Shelf,* 35 BRIT. Y.B. INT'L L. 102, 120 (1959) (footnote omitted). *Accord,* Lauterpacht, *Sovereignty over Submarine Areas,* 27 BRIT. Y.B. INT'L L. 376, 410 (1950); Padwa, *Submarine Boundaries,* 9 INT'L & COMP. L.Q. 628, 644 (1960); M. McDOUGAL & W. BURKE, THE PUBLIC ORDER OF THE OCEANS 436–37 (1962).

The general problem of small islands has been discussed in more detail elsewhere.[75]

In the South China Sea, because of the great number of very small islets far from the major coasts, many of them subject to conflicting claims of ownership, the problem is particularly vexing.[76] Furthermore, note must be taken of another special circumstance in this part of the world, the existence of deep troughs in the seabed.

Seabed Boundaries in the South China Sea

Agreed Lines

Few South China Sea final boundary agreements have been published, and those boundaries that have been made final are lateral ones. In 1965 the British Government published orders establishing certain of the boundaries of the then territories of Brunei and Sarawak in North Borneo. The successor governments are understood to have adhered to these boundary determinations. In 1969 the Federation of Malaysia and the Republic of Indonesia agreed on two continental shelf boundary lines extending out into the South China Sea from their adjacent land territories.[77] This agreement is indicated on Map 6.

Application of the Equidistance Principle

The methods of seabed boundary demarcation prescribed in Article 6 of the Continental Shelf Convention, where the equidistance principle is subject to modification to accommodate special circumstances, are of course binding only between parties to the Convention.[78] In the South China Sea only the Republic of China (Taiwan province) and the Federation of Malaysia are parties. Nevertheless, the methods provided in Article 6 combine to generate a quality that will be highly valued by diplomats seeking agreement on national seabed boundaries in the South China Sea: flexibility. Though the equidistance principle

75. Ely, *Seabed Boundaries Between Coastal States: The Effect to be Given Islets as "Special Circumstances,"* 6 INT'L LAWYER 219 (1972).

76. *See* note 52 *supra.*

77. *See* Office of the Geographer, Bureau of Intelligence & Research, U.S. Dep't of State, International Boundary Study, Ser. A: Limits in the Seas, Continental Shelf Boundary, Indonesia-Malaysia, No. 1 (1970). This publication shows the straight base lines established by the Federation of Malaysia.

78. [1969] I.C.J. 3, 38–39 (para. 38).

of Article 6 has not attained the status of customary international law with respect to lateral boundaries, and may not have done so as to frontal boundaries, it has been the basis for bilateral seabed agreements in many parts of the world[79] and would be a useful focus for seabed boundary demarcation techniques in the South China Sea. Moreover, the reasons given by the Court for refusing to apply that principle with respect to Germany's lateral boundaries with her neighbors are not present in the South China Sea, save in the single case of Brunei.

The device of the median line, applied between States whose coasts are opposite each other, enables us to construct sufficient additional lines of seabed boundary demarcation to complete a generalized prima facie estimate of national seabed jurisdiction in the South China Sea. Median lines are shown on Map 6. A principal median line should be constructed between the Asian coast States—the Republic of Vietnam, and the People's Republic of China, and the Asian island arc States—the Republic of China (Taiwan province), the Republic of the Philippines, the Federation of Malaysia, Brunei, and the Republic of Indonesia. Three shorter median lines should then be drawn extending out into the South China Sea to meet the principal line. These three lines separate the opposite seabed jurisdictions of the Republic of China (Taiwan province) and the Republic of the Philippines, of the Republic of the Philippines and the Federation of Malaysia, and of the Republic of Vietnam and the People's Republic of China.[80] The use of equidistance lines in determining lateral boundaries does not produce the inequitable results that caused the Court to reject it in the *North Sea Continental Shelf Cases,* with the exception of the boundaries of Brunei, which should be determined like those of Germany.

79. Office of the Geographer, Bureau of Intelligence & Research, U.S. Dep't of State, International Boundary Study, Ser. A: Limits in the Seas, Nos. 1, 2, 9, 12, 16, 17, 18, 24, 25, 26 (1970). Between Jan. 21, 1970, and Oct. 19, 1973, 56 items in this valuable series have been published.

80. The writers have chosen to treat these three lines as median lines between opposite coasts across the Luzon and Balabac Straits and the Tonkin Gulf, respectively. The International Court of Justice has distinguished carefully between application of the median line as between opposite States, which it impliedly approves as customary international law, and the principle of equidistance as between adjacent States, which it does not approve. [1969] I.C.J. 3, 37–38 (paras. 57 and 58), as noted *supra*. Were one, or all, of these lines to be considered a line between adjacent coasts (say, for example, the Republic of the Philippines/Federation of Malaysia median line), it would be a lateral equidistance line, not a median line, in the Court's usage. Lateral equidistance lines are much more difficult to construct because they tend to leave "to one of the States concerned areas that are a natural prolongation of the territory of the other." *Id.* at 38 (para. 58).

Special Circumstances

The effects of three types of special circumstances in the South China Sea were examined before seabed boundaries were constructed.[81] The physical features that might be considered special circumstances affecting the extent of national seabed jurisdiction are, first, the presence of islands, islets, and rocks; second, the depths of troughs and trenches; and third, the shapes and lengths of coastlines.

Islets

The South China Sea is riddled with hundreds of islands, islets, and rocks, and, just beneath the water surface, underlain by many reefs and shoals. Because islands have territorial seas,[82] and because Article 6 of the Continental Shelf Convention draws median lines of seabed jurisdiction from the base lines of the territorial sea, the question arises whether to use islands to construct median lines in the South China Sea.

The fact that an island's base line must be recognized for all other purposes[83] but may be denied recognition for the measurement of continental shelf boundaries need not concern us, for the consequences of the recognitions are vastly different. The area of the continental shelf that may be affected by recognition or nonrecognition of an island's base lines for calculation of a seabed boundary may be hundreds, or indeed thousands, of times as great as the area of its territorial sea or contiguous zone, measured from those same base lines.

81. Special circumstances are a method of seabed boundary demarcation noted in Article 6 of the Continental Shelf Convention, *supra* note 1.
82. The Convention on the Territorial Sea and the Contiguous Zone, *done* Apr. 29, 1958, [1964] 2 U.S.T. 1606, T.I.A.S. No. 5639, 516 U.N.T.S. 205 (in force for U.S. September 10, 1964) [hereinafter cited as Territorial Sea Convention]. Article 10.1 of the Convention defines an island as a "naturally-formed area of land, surrounded by water, which is above water at high-tide." There is no size restriction here. Article 3 provides that "[e]xcept where otherwise provided in these articles, the normal baseline for measuring the breadth of the territorial sea is the low-water line along the coast as marked on large-scale charts officially recognized by the coastal State." Article 10.2 provides that "[t]he territorial sea of an island is measured in accordance with the provisions of these articles." Thus, islets, of whatever size, have base lines and territorial seas measured from those base lines. Islets also have contiguous zones for customs, fiscal, immigration, and sanitary purposes under Article 24 of the same Convention; the contiguous zone may not extend "beyond twelve miles from the baseline from which the breadth of the territorial sea is measured." For the importance of median lines to territorial seas and contiguous zones, see Articles 12 and 24.3.
83. *Id.*

Simple distinctions are not easily made between islands important enough to be recognized as base lines for the drawing of median lines as against the coast of the mainland or a major island, and islets that should be denied such recognition because of the inequitable distortion of seabed boundaries that their recognition would occasion.[84] To give effect to all the South China Sea islands, islets, and rocks in constructing median lines well might be a task beyond the capabilities of modern cartographic science. Once plotted, such a honeycomb of median lines would attract an even more complicated legal debate over the ownership of the islands. For this reason, subject always to agreement between the States concerned, the writers choose to ignore the effect on median lines of all islands lying beyond twenty-four nautical miles (the sum of the two contiguous territorial seas of twelve miles) of the low-water line on the coast of the mainland or of a major island such as Palawan or Hainan.[85] Beyond twenty-four nautical miles, we doubt the efficacy of any criterion of population, size, military, commercial, or navigational significance, distortion, or equity to distinguish among the hundreds of midsea islands and islets in the South China Sea. Accordingly, these writers ignore them all. The burden of proof is on any State or other party choosing to assert such islets as base lines for seabed boundary median lines. The authors would accord to each of these hundreds of small islands in the South China Sea a portion of the continental shelf coterminous with its twelve-mile contiguous zone.[86] The special circumstances of these

84. The legal literature on islands is profuse. *See,* R. D. HODGSON, ISLANDS: NORMAL AND SPECIAL CIRCUMSTANCES (1973, Research Study, The Geographer, Bureau of Intelligence and Research, U.S. Dep't of State). *See also* Gutteridge, Lauterpacht, Padwa, McDougal and Burke, *supra* note 74. *See also* Oda, *Boundary of the Continental Shelf,* 12 JAPANESE ANN. INT'L L. 264, 280–83 (1968); Kennedy, Brief Remarks on Median Lines and Lines of Equidistance and on the Methods Used in Their Construction 7–8, paper distributed by the United Kingdom Delegation to the Conference on the Law of the Sea, Apr. 2, 1958); 4 M. WHITEMAN, DIGEST OF INTERNATIONAL LAW 912, 913 (1965); Boggs, *Delimitation of Seaward Areas Under National Jurisdiction,* 45 AM. J. INT'L L. 240 (1951).
85. As a variation of the straight base lines principle of the Norwegian Fisheries Case, [1951] I.C.J. 116, of Article 4 of the Territorial Sea Convention, note 82 *supra,* and of the North Sea Continental Shelf Cases, [1969] I.C.J. 3, 52 (para. 98), the present writers suggest that a South China Sea islet may be used as a base line for seabed boundary demarcation if any portion of the islet lies within 24 nautical miles of the coast of its owner's mainland or major island. This islet's 12-mile contiguous zone merges with the 12-mile contiguous zone of the larger land territory, forming an envelope encompassing both. *See* Ely, *supra* note 75.
86. The connection is persuasive between the State's limited jurisdiction in the contiguous zone and its limited jurisdiction with respect to the resources of the continental shelf, the right in both cases being one which may extend seaward of the territorial sea. They are legal "concepts of the same kind." [1969] I.C.J. 3, 51 (para. 96).

islets can be recognized by holding that other States cannot extend their continental shelf rights to explore and exploit submarine resources into the seabed and subsoil underlying the waters of the islet's contiguous zone. Only the State owning that islet should be recognized as having jurisdiction with respect to the seabed resources within that zone. Because of its scale, Map 6 does not indicate such circumferential zones of continental shelf rights for these hundreds of islands and islets.

In the case of archipelagos, or clusters of islands, this island seabed jurisdiction might be expanded to permit straight lines to be drawn from one island to another, attributing to each such archipelago a share of the continental shelf encompassed not only by the several islands' contiguous zones but also by the straight base lines, plus a twelve-mile belt measured from those lines.[87] But, again, this conglomerate should not be recognized as a base line for drawing median lines, save as against an islet or similar cluster whose contiguous zone, similarly enlarged, impinges on that of the first cluster.

Thus, the writers conclude that islands, islets, and rocks in the South China Sea are special circumstances and therefore should not be recognized as generating base lines in the demarcation of national seabed boundaries.

Troughs

The principal trough formations in the South China Sea are the Palawan Trough, the West Luzon Trough and the Manila Trench. Each of these geomorphological depressions is found off of the western coast of the Republic of the Philippines. Does the presence of these troughs so insulate the Philippines island arc from the seabed of the South China Sea as to constitute a special circumstance that deprives the Republic of the Philippines of the right to use its western coast as a base line for the demarcation of a median line as against the Asian mainland? These writers think not. If our basic conclusion is sound—that the Asian continental margin ends in the deep trenches on the Pacific side of the Philippines, not in these South China Sea troughs, and that the Philippines are among the island arcs that en-

---

87. Some archipelago claims are quite unrelated to the straight base line concept, and go much further. Archipelago claims have not been recognized by treaty, so far as we have found, except to the limited extent of the Federation of Malaysia / Republic of Indonesia agreement, and perhaps in the case of the Republic of the Philippines, *supra* note 68. *See* R. Hodgson & L. Alexander, Towards an Objective Analysis of Special Circumstances 45 (1972).

close the South China Sea as part of the Asian continental margin—it follows that the Palawan Trough, the West Luzon Trough and the Manila Trench have no legal significance in the determination of seabed boundaries. On Map 6 the seabed jurisdiction of the Republic of the Philippines jumps these troughs and meets the jurisdiction of Asian mainland States at a median line in the South China Sea.

The Special Case of Brunei

The shape of the coastline of northwestern Borneo should be considered a special circumstance insofar as its concavity, running southwest along the coastlines of the Malaysian state of Sabah, Brunei and the Malaysian state of Sarawak, denies an equitable share of seabed jurisdiction to Brunei. This conclusion is in accord with the *North Sea Continental Shelf Cases.*[88] There the Court declared principles that, when later applied,[89] allowed the Federal Republic of Germany to divert its seabed jurisdiction out to the North Sea intercoastal median line and avoid a pinching-off of its seabed area due to the presence of adjacent States on a concave coast. The Court set forth several equitable factors, as previously discussed, to govern the drawing of equidistance lines that apply as well to the situation of Brunei in the South China Sea.[90] The equidistance-line boundaries of Brunei as against the Federation of Malaysia would pinch off Brunei's share of the seabed far short of the median line with the Republic of Vietnam. Avoiding this inequity, Map 6 follows the teaching of the North Sea cases and extends Brunei's seabed area to the Asian coast–Asian island arc median line.

National Coastline Lengths

The coastal States of the South China Sea compare well with the rest of the world as to coastline length. Statistics for the world's coastlines reveal that several of the seven South China Sea littoral States are

---

88. Giving particular attention to concavity or convexity of coastlines, and their relative lengths, as equitable factors to be taken into account in weighing the applicability of the equidistance principle, the Court sought "to establish the necessary balance between States with straight, and those with markedly concave or convex coasts . . . ." [1969] I.C.J. 3, 52 (para. 98).
89. 10 INT'L LEGAL MATERIALS 600–12 (1971) (*see* map at 602).
90. *See* text accompanying note 72 *supra.*

long-coastline States.[91] Of course, the fact that the Republic of the Philippines and the Republic of Indonesia are island nations accounts for their huge coastlines. Nevertheless, if the "lengths of . . . coastlines"[92] are an equitable consideration in North Sea seabed boundary delimitation, they well might be so also in the South China Sea. Three of the ten nations of the world with the longest coastlines are South China Sea littoral States.[93] Clearly, the South China Sea coastal States have great interest in the use of coastline lengths as a criterion of the extent of national seabed resource jurisdiction. Insofar as their coastlines adjoining the South China Sea are concerned, however, it would appear from Map 6 that the equities of these littoral States are reflected in the median and equidistance lines there proposed.

The Patrimonial Sea Concept

This emerging diplomatic doctrine is particularly appropriate in the semienclosed or marginal seas. Indeed, Ecuador, a nonparticipant in the June 1972 Santo Domingo conference, has commented that "[t]he statement by the representative of Venezuela introducing the Declaration of Santo Domingo had made it very clear that the Declaration was the outcome of a specialized conference of largely Caribbean countries and did not reflect the viewpoint of Latin America. . . . [T]he Declaration was in harmony with the geographical, political and sociological characteristics of a practically closed sea containing highly differentiated insular States, living in close geographical proximity."[94]

91. A national list, ranked in order of coastline size, includes:

Order	Political Entity	Miles of coastline	% of total of world
2	Republic of Indonesia	19,784	10.89
6	Republic of the Philippines	6,997	3.85
10	People's Republic of China	3,492	1.92
20	Federation of Malaysia	1,853	1.02
42	Republic of Vietnam	865	0.48
60	Republic of China (Taiwan province)	470	0.26
Small-coast territory	Brunei	88	0.05

This list was compiled from data from Office of the Geographer, Bureau of Intelligence & Research, U.S. Dep't of State, Geographic Bull. No. 3, Sovereignty of the Sea, Table II, at 18 *et seq.* (rev. Oct. 1969).
92. [1969] I.C.J. 3, 52 (para. 98).
93. *See* note 91 *supra.*
94. U.N. Doc. A/AC.138/SR.80, at 2 (Jul. 28, 1972).

Under this analysis, if the patrimonial sea concept is appropriate for the Caribbean Sea, it surely is appropriate for the South China Sea. The effect of such a 200-mile line is shown on Map 6.

Conclusion

It is these writers' opinion, that the States littoral to the South China Sea and these States alone, under existing principles of customary international law, have sovereign rights with respect to the exploration and exploitation of the natural resources of the seabed and subsoil of that sea. Their respective national seabed jurisdictions might conform generally to the boundaries on Map 6. The writers do not assert that this conclusion is immune to counterargument. There is too little law of a controlling character to support a dogmatic position on the extent of national seabed jurisdiction in this difficult area.

Appendix 1

Map 5 is merely a location map, indicating the relationship of the string of Asian semienclosed seas to the eastern edge and southeastern corner of the Asian continent.

Map 6 (in pocket at back of book) is an attempt to construct approximate South China Sea seabed boundaries in accordance with existing principles of customary international law. The writers must emphasize the tentative, approximate nature of these lines. They are useful for illustration only. The coordinates of the turning points of these lines have been rounded off to the nearest five minutes of longitude and latitude, partly because this eased our cartographers' burden and had little distorting effect on our large-scale map, but principally to make sure that the map might not be mistaken for a definitive treatment of South China Sea seabed jurisdiction. Given the huge area of the subject, such perfect science was beyond both our capabilities and intentions.

The base instrument of Map 6 is the British Admiralty Map No. 1263, modified for our purpose. First compiled in 1886, No. 1263 was revised in 1966. Small corrections were made through 1972. The unnumbered, small-dash lines following the coasts of the Republic of the Philippines, the Federation of Malaysia, the Republic of Indonesia and the People's Republic of China are straight coastal base lines.[95] The

95. Office of the Geographer, Bureau of Intelligence & Research, U.S. Dep't of State, International Boundary Study, Ser. A: Limits in the Seas, Straight Baselines, The Philippines, No. 33 (1971); Agreement Between the Government of the Republic of Indonesia and the Government of Malaysia Relating to the Delimitation of the Continental Shelves Between the Two Countries, Oct. 27, 1969, 9 INT'L LEGAL MATERIALS 1173 (1970); Office of the Geographer, Bureau of Intelligence & Research, U.S. Dep't of State, International Boundary Study, Ser. A: Limits in the Seas, Straight Baselines, Indonesia, No. 35 (1971), *also found at* U.N. Doc. A/ CONF. 19/5/Add. 1 (Apr. 4, 1960); Declaration on China's Territorial Sea, Sept. 4, 1958, 1 PEKING REV., Sept. 9, 1958, No. 28, at 21; Office of the Geographer, Bureau of Intelligence & Research, U.S. Dep't of State, International Boundary Study, Ser. A: Limits in the Seas, Straight Baselines, People's Republic of China, No. 43 (1972). This last study maps a hypothetical system of Chinese base lines from available documents, which include: English language broadcasts from Peking, Peking New China News Agency International Service, Dec. 3, 24, 29, 31, 1970; Wash. Post, Dec. 5, 1970, §A, at 1, col. 1. *See also* Speech by An Chih-Yuan,

large-dash line about the Republic of the Philippines is that State's proclaimed boundary according to the 1898 Spain–United States treaty.[96] The oblong line with shaded edge in the center of the South China Sea represents the limit of a coastal State 200-mile resource zone, or patrimonial sea.[97] The dotted line offshore of the Republic of Vietnam is that State's proclaimed continental shelf line.[98] As stated above, the authors have chosen not to indicate zones of continental shelf rights for the hundreds of South China Sea islands and islets. Lastly, the equidistance lines ignore the effect of all islands that lie beyond 24 nautical miles of the low-water line on the coast of the mainland or a major island, such as Palawan or Hainan.

Line *A* is a major intercoastal median line between the Asian coast States and the Asian island arc States. Line *B* is a median line between the Republic of China (Taiwan province) and the Republic of the Philippines. Line *C* is a median line between the Republic of the Philippines and the Federation of Malaysia. Line *D* is an estimated lateral equidistance line between the adjacent States of the Federation of Malaysia and Brunei. Line *E* is also an estimated lateral equidistance line between these two adjacent States. Line *F* is the eastern continental shelf boundary agreed between the Federation of Malaysia and the Republic of Indonesia. Line *Z* arbitrarily extends Line *F* to link up Line *F* and Line *A*. Line *G* is the western continental shelf boundary agreed between the Republic of Indonesia and the Federation of Malaysia. Line *Z*[1] arbitrarily extends Line *G* to link up Line *G* and Line *A*. Line *H* is a median line between the Republic of Vietnam and the People's Republic of China. See Appendix 2 for turning point coordinates.

Representative of the People's Republic of China, at the U.N. Comm. on the Peaceful Uses of the Sea-Bed and the Ocean Floor Beyond the Limits of National Jurisdiction, Mar. 3, 1972, New York, 11 INT'L LEGAL MATERIALS 654 (1972). For an extensive bibliography of Chinese physical oceanography and marine geology, see Wong & Ku, *Oceanography and Limnology in Mainland China,* U.N. ECON. COMM'N FOR ASIA & THE FAR EAST, COMM. FOR COORDINATION OF JOINT PROSPECTING FOR MINERAL RESOURCES IN ASIAN OFFSHORE AREAS, 3 TECHNICAL BULL. 137 (1970). *See also* Cheng, *Communist China and the Law of the Sea,* 63 AM. J. INT'L L. 47 (1969).

96. Treaty of Peace Between the United States and Spain, Dec. 10, 1898, 32 *Martens Nouveau Recueil* 74, ser. 2 (1905). The 1898 line is indicated on Map 6 in pocket part.

97. *Specialized Conference of the Caribbean Countries Concerning the Problems of the Sea: Declaration of Santo Domingo,* June 7, 1972, U.N. Doc. A/AC. 138/80, at 6 (Jul. 26, 1972), *reproduced in* 11 INT'L LEGAL MATERIALS 892 (1972).

98. Decree No. 81/NG of Apr. 27, 1965, *reproduced in* 4 INT'L LEGAL MATERIALS 461 (1965); Decree No. 249/BKT/VP/UBQGDH/ND of June 9, 1971, found in a letter from the United Kingdom Foreign and Commonwealth Office, Nov. 28, 1972.

Appendix 2

References are to Map 6 (pocket part)

Line A: *Asian coast–Asian island arc median line*

Points	People's Republic of China / Republic of China (Taiwan province)		
A 1	118° 15' E.	22° 30' N.	(Formosa Strait arbitrary limit)
A 2	118° 10' E.	21° 05' N.	
A 3	118° 00' E.	20° 20' N.	(Taiwan, Philippines, China, tri-section point)
	People's Republic of China / Republic of the Philippines		
A 4	117° 20' E.	19° 45' N.	
A 5	116° 50' E.	19° 10' N.	
A 6	116° 30' E.	18° 30' N.	
A 7	116° 15' E.	18° 30' N.	
A 8	116° 10' E.	18° 25' N.	
A 9	115° 20' E.	17° 40' N.	
A 10	114° 45' E.	16° 20' N.	
A 11	114° 30' E.	15° 20' N.	(Philippines, Vietnam, China, trisection point)
	Republic of Vietnam / Republic of the Philippines		
A 12	114° 40' E.	14° 30' N.	
A 13	114° 45' E.	14° 00' N.	
A 14	114° 25' E.	12° 55' N.	
A 15	114° 10' E.	12° 30' N.	
A 16	113° 35' E.	11° 25' N.	
A 17	113° 10' E.	10° 20' N.	
A 18	112° 45' E.	09° 15' N.	
A 19	112° 35' E.	08° 55' N.	(Philippines, Malaysia, Vietnam, trisection point)
	Republic of Vietnam / Federation of Malaysia		
A 20	112° 30' E.	08° 40' N.	
A 21	112° 20' E.	08° 40' N.	(Malaysia, Brunei, Vietnam, trisection point)

Points

A 22	111° 05' E.	07° 45' N.	
A 23	110° 45' E.	07° 45' N.	
A 24	110° 25' E.	07° 45' N.	
A 25	109° 11' E.	07° 23' N.	(Malaysia, Indonesia, Vietnam, projected intersection)

Republic of Vietnam / Republic of Indonesia

A 26	105° 60' E.	06° 15' N.	
A 27	105° 50' E.	06° 15' N.	(Indonesia, Malaysia, Vietnam, projected intersection)

Republic of Vietnam / Federation of Malaysia

A 28	105° 10' E.	06° 25' N.	
A 29	104° 35' E.	06° 40' N.	
A 30	103° 25' E.	07° 30' N.	(Gulf of Thailand arbitrary limit)

Line B: *Republic of China (Taiwan province) / Republic of the Philippines median line*

B 1	121° 45' E.	21° 30' N.	
B 2	121° 20' E.	21° 20' N.	
B 3	120° 45' E.	20° 50' N.	
B 4	120° 20' E.	20° 35' N.	
B 5	120° 10' E.	20° 25' N.	
B 6	119° 10' E.	20° 15' N.	
A 3	118° 00' E.	20° 20' N.	(Taiwan, Philippines, China, trisection point)

Line C: *Republic of the Philippines / Federation of Malaysia median line*

C 1	116° 40' E.	07° 15' N.	
C 2	115° 05' E.	08° 10' N.	
A 19	112° 35' E.	08° 55' N.	(Philippines, Malaysia, Vietnam, trisection point)

Line D: *Estimated Federation of Malaysia / Brunei eastern lateral seabed boundary*

D 1	115° 10' E.	05° 10' N.	
A 21	112° 20' E.	08° 40' N.	(Malaysia, Brunei, Vietnam, trisection point)

Line E: *Estimated Brunei / Federation of Malaysia western lateral seabed boundary*

E 1	114° 10' E.	04° 35' N.	
A 21	112° 20' E.	08° 40' N.	(Malaysia, Brunei, Vietnam, trisection point)

Points

Line F: *Agreed Federation of Malaysia / Republic of Indonesia continental shelf eastern boundary*

F 1	109° 38.8' E.	02° 05' N.
F 2	109° 54.5' E.	03° 00' N.
F 3	110° 02' E.	04° 40' N.
F 4	109° 59' E.	05° 31.2' N.
F 5	109° 38.6' E.	06° 18.2' N.

Line Z: *Arbitrary connection of Line F with Line A*

F 5	109° 38.6' E.	06° 18.2' N.	
A 25	109° 11' E.	07° 23' N.	(Malaysia, Indonesia, Vietnam, projected intersection)

Line G: *Agreed Republic of Indonesia / Federation of Malaysia continental shelf western boundary*

G 1	104° 29.5' E.	01° 23.9' N.
G 2	104° 53' E.	01° 38' N.
G 3	105° 05.2' E.	01° 54.4' N.
G 4	105° 01.2' E.	02° 22.5' N.
G 5	104° 51.5' E.	02° 55.2' N.
G 6	104° 46.5' E.	03° 50.1' N.
G 7	104° 51.9' E.	04° 03' N.
G 8	105° 28.8' E.	05° 04.7' N.
G 9	105° 47.1' E.	05° 04.7' N.
G 10	105° 49.2' E.	06° 05.8' N.

Line Z[1]: *Arbitrary connection of Line G with Line A*

G 10	105° 49.2' E.	06° 05.8' N.	
A 27	105° 50' E.	06° 15' N.	(Indonesia, Malaysia, Vietnam, projected intersection)

Line H: *Republic of Vietnam / People's Republic of China median line*

H 1	108° 10' E.	17° 35' N.
H 2	108° 25' E.	17° 20' N.
H 3	108° 40' E.	17° 15' N.
H 4	108° 60' E.	17° 10' N.
H 5	109° 15' E.	17° 00' N.
H 6	109° 55' E.	16° 45' N.
H 7	110° 25' E.	16° 40' N.
H 8	111° 05' E.	16° 30' N.

IV

National Seabed Boundary Options

Robert D. Hodgson and Terry V. McIntyre

Although the basis for a national selection of a seabed boundary should be reason, few States possess the knowledge and the understanding upon which to base a reasonable claim. As a result, expansionist claims for national rights to the seabed tend to become more and more commonplace. Not surprisingly, common heritage, it would seem, has become "mine," not "ours." Great expectations of wealth from the seabed have been raised, not by diplomats and politicians of the developing countries, but by scientists and technicians of the developed world. The vision of wealth for the taking results in a manifest destiny approach to resources in adjacent waters. The poor are prone to think of the common heritage of mankind as a panacea to alleviate their poverty. Why should the developing coastal State vitally in need of more revenues seek to have these resources allocated to an international organization subject to the manipulation of the resource-hungry developed States?

To choose the reasonable option becomes difficult. The physical properties of the oceans make study difficult and expensive. Sophisticated instruments must be tended by trained scientists to extract knowledge of the sea and its resources. Costs are high, and few countries can risk the expense involved in diving the deep. The oceans, however, represent the last frontier on this planet. All fruitful lands have been occupied, all islands claimed. Only the ice-covered Antarctic and the open oceans hold any promise for the extension of a national economic base for development. Inevitably, the seas also will be exploited.

Two channels for development of the oceans are open: an internationalization of the seabed or the spread of national domain over the ocean floor. States take sides, espousing one principle or the other. But ignorance is the worst enemy of a rational selection of the limits between national and international zones of the seabed. Uninformed about the resource location, States may be tempted to demand the

This chapter does not represent official United States policy concerning the positions or attitudes expressed but reflects only the positions of the authors.

greatest possible national extension in the hope that resource location will equate with area. That hope appears doomed. Resources on land are distributed unequally, and it is likely that the same condition will prevail under the seas. It is unfortunate that States in search of petroleum should demand additional seabed area to compensate for narrow continental shelves, especially when one considers the plight of land-locked States. If, as it appears, most of the world's exploitable petroleum is to be found on the continental shelf and slope, to what avail is the annexation of deep seabed beyond these limits?

We do not know, of course, what resources the seabed may contain, but surely the greater this depth and distance from the shore, the higher will be the cost of exploration and exploitation. Deep seabed deposits will thus have to increase geometrically in extent or magnitude in order to warrant investment. Moreover, the greater the extent of national claims, the less likely the international domain will contain resources of value for the common heritage of mankind.

The following limits to national jurisdiction have been advanced:

1. Two hundred meters, modified by adjacency and exploitability, as contained in the current Convention on the Continental Shelf,[1] which forms a part of the United States Draft[2]

2. Two hundred meters depth or 40-nautical miles distance, being the formula advanced in the United Nations Seabed Committee by the Seven Power Land- and Shelf-Locking States Working Paper[3]

3. Two hundred nautical miles as claimed by certain States as a limit of the national territorial sea or as a seaward limit of a patrimonial sea[4]

4. The seaward edge of the margin that marks the boundary between the United States–proposed trusteeship and truly international zones[5]

The Office of the Geographer in the United States Department of State has plotted on small-scale charts the national limits to the seabed according to the various proposals that have been placed before the United Nations Seabed Committee or have been advanced

---

1. Convention on the Continental Shelf, *done* Apr. 28, 1958, [1964] 1 U.S.T. 471, T.I.A.S. No. 5578, 499 U.N.T.S. 311.
2. 65 DEP'T STATE BULL. 1680 (1971), 10 INT'L LEGAL MATERIALS 1013.
3. Proposal of the Seven Powers, U.N. Doc. A/AC. 138/55 (1970). *See also* Seven Power Working Paper, U.N. Doc. A/8421 (1970).
4. For claims of these and other nations, see generally Office of the Geographer, Bureau of Intelligence & Research, U.S. Dep't of State, International Boundary Study, Ser. A: Limits in the Sea, National Claims to Maritime Jurisdiction, No. 36 (1972).
5. 9 INT'L LEGAL MATERIALS 1046 (1970).

internationally. The resulting areas have been measured by computer or planimeter. The results of this study are included as Tables 2–4.[6]

From analyzing these data, we arrive at certain options for the world's nations in negotiating the limits between their national jurisdictions and an international zone. Obviously, no one is privy to the many factors that may affect a nation's evaluation of the options that could lead to the ultimate policy selection. Consequently, these options, although they may occasionally be expressed rather positively, may not predominate in national thinking. One need only examine the various factors relating to the United States policy decisions to note the complexity of the process. Military and commercial shipping call for narrow national territorial sea limits. They do not demand the same restrictions for the seabed, but they share a fear of creeping jurisdiction—the nagging fear that broad seabed limits will, at some later period, be expanded to full sovereign demands. To thwart this possibility, which may not be a remote one, these interests feel compelled to press also for a limited national domain on the sea floor for fear of the indirect effects of these demands.

Fisheries interests have proponents on both sides of the boundary option fence. Domestic fishermen prefer an extended and exclusive economic resource zone that would restrict foreign distant-water fishermen. United States tuna and shrimp fleets, however, are appalled at this thought. Petroleum companies generally favor wide national jurisdiction, but again the interests are not universal. On the other hand, some government officials argue for a narrow shelf in the hope that the revenues from a large international seabed area will aid in the economic development of the less-developed countries. Internationalists seek the same goal but for differing purposes. They believe that an internationalization of the seabed will strengthen the cause of world government, the United Nations, and other supranational organizations.

How these competing and conflicting viewpoints will affect the final policy is difficult to predict. On which factor does the State place greatest emphasis: international strategic implications, local fishing, distant-water fishing, worldwide aviation, internationalism, economic development, commercial shipping, environmental protection? Plainly, in a democratic country such as the United States, the ultimate national policy will reflect a compromise—a balancing of interests to protect general aims without compromising the specific needs.

6. *Reprinted from* Office of the Geographer, Bureau of Intelligence & Research, U.S. Dep't of State, International Boundary Study, Ser. A: Limits in the Seas, Theoretical Areal Allocations of the Seabed to Coastal States Based on Certain U.N. Seabed Committee Proposals, No. 46 (1972).

Table 2. Rank order of coastal States by area and by length of coast

Country area	Coastal length
1. Soviet Union	Soviet Union
2. Canada	Indonesia
3. China, People's Republic of	Australia
4. United States	United States
5. Brazil	Canada
6. Australia	Philippines
7. India	Japan
8. Argentina	Mexico
9. Sudan	Brazil
10. Algeria	China, People's Republic of
11. Zaire	Chile
12. Saudi Arabia	India
13. Mexico	New Zealand
14. Indonesia	United Kingdom
15. Libya	Italy
16. Iran	Spain
17. Peru	Madagascar
18. Ethiopia	Argentina
19. South Africa	Malaysia
20. Colombia	Turkey
21. Mauritania	Cuba
22. Egypt	Norway
23. Tanzania	Greece
24. Nigeria	Somalia
25. Venezuela	South Africa
26. South-West Africa	France
27. Pakistan	Sweden
28. Turkey	Egypt
29. Chile	Peru
30. Burma	Saudi Arabia
31. Somalia	Thailand
32. Madagascar	Burma
33. Kenya	Iceland
34. France	Venezuela
35. Thailand	Colombia
36. Spain	Oman
37. Cameroon	Panama
38. Sweden	Iran
39. Morocco	Libya
40. Iraq	Morocco
41. Japan	South Vietnam
42. Congo	Sri Lanka (Ceylon)
43. Finland	Denmark

Table 2. Continued

Country area	Coastal length
44. Malaysia	Finland
45. Ivory Coast	Ireland
46. Norway	Korea, South
47. Poland	Portugal
48. Italy	Yemen (Aden)
49. Philippines	South-West Africa
50. Yemen (Aden)	Tanzania
51. Ecuador	Algeria
52. New Zealand	Haiti
53. Gabon	Korea, North
54. Yugoslavia	Tunisia
55. Germany, Fed. Rep. of	China, Republic of
56. Guinea	Costa Rica
57. United Kingdom	Ecuador
58. Ghana	Gabon
59. Romania	Honduras
60. Guiana	Mauritania
61. Oman	Nicaragua
62. Senegal	Nigeria
63. Yemen (Sana)	Pakistan
64. Syria	Sudan
65. Cambodia	United Arab Emirates
66. Uruguay	Vietnam, North
67. Vietnam, South	Yugoslavia
68. Tunisia	Bangladesh
69. Vietnam, North	Cyprus
70. Nicaragua	Dominican Republic
71. Bangladesh	Ethiopia
72. Greece	Germany, Fed. Rep. of
73. Korea, North	Ghana
74. Cuba	Ivory Coast
75. Bulgaria	Jamaica
76. Dahomey	Liberia
77. Honduras	Trinidad
78. Liberia	Uruguay
79. Guatemala	Albania
80. German Democratic Republic	Cambodia
81. Iceland	Cameroon
82. Korea, South	El Salvador
83. Jordan	Equatorial Guinea
84. Portugal	German Democratic Republic
85. United Arab Emirates	Guatemala
86. Panama	Guinea

Table 2. Continued

Country area	Coastal length
87. Sierra Leone	Guyana
88. Ireland	Kenya
89. Sri Lanka (Ceylon)	Netherlands
90. Togo	Poland
91. Costa Rica	Qatar
92. Dominican Republic	Senegal
93. Denmark	Sierra Leone
94. China, Republic of	Yemen (Sana)
95. Netherlands	Bulgaria
96. Belgium	Israel
97. Albania	Kuwait
98. Equatorial Guinea	Romania
99. Haiti	Lebanon
100. El Salvador	Fiji
101. Israel	Mauritius
102. Kuwait	Maldives
103. Qatar	Tonga
104. Cyprus	Congo
105. Jamaica	Syria
106. Gambia	Bahrain
107. Lebanon	Dahomey
108. Fiji	Barbados
109. Trinidad	Malta
110. Western Samoa	Gambia
111. Mauritius	Belgium
112. Tonga	Togo
113. Bahrain	Singapore
114. Singapore	Western Samoa
115. Barbados	Zaire
116. Maldives	Nauru
117. Malta	Iraq
118. Nauru	Jordan

If the majority of the world's coastal States determine that the common heritage of mankind equates with the maximum area under national jurisdiction, there can be no doubt but that a 200–nautical mile zone would be the choice. A 200-mile zone would appropriate the greatest national seabed area: 37,745,000 square nautical miles (35.86 percent of the world's ocean beds), which would be the most important and most easily exploitable in terms of ocean resources. The residual area, although extensive, would probably yield only the hard minerals of the manganese nodules. Beyond 200 nautical miles,

Table 3. National ranking based on sum of ranks for seabed proposals

1. Indonesia	47. Morocco
2. Canada	48. Panama
3. Australia	49. Pakistan
4. United States	50. Sri Lanka (Ceylon)
5. Soviet Union	51. Jamaica
6. Brazil	52. Honduras
7. Japan	53. Finland
8. Mexico	54. Nicaragua
9. New Zealand	55. Tanzania
10. India	56. Egypt
11. Argentina	57. Gabon
12. China, People's Republic of	58. Nigeria
13. Norway	59. Tonga
14. United Kingdom	60. Maldives
15. Spain	61. Dominican Republic
16. Madagascar	62. Netherlands
17. Philippines	63. North Vietnam
18. Italy	64. Fiji
19. Iceland	65. Tunisia
20. South Africa	66. North Korea
21. South Vietnam	67. Uruguay
22. Malaysia	68. Mauritania
23. Burma	69. Ethiopia
24. Chile	70. Guiana
25. Mauritius	71. Costa Rica
26. Thailand	72. Senegal
27. Korea, South	73. United Arab Emirates
28. Somalia	74. Denmark
29. Peru	75. Ghana
30. Colombia	76. Liberia
31. Cuba	77. Sudan
32. France	78. Equatorial Guinea
33. Ecuador	79. Bangladesh
34. Venezuela	80. Cambodia
35. Ireland	81. Algeria
36. Portugal	82. Haiti
37. Yemen (Aden)	83. Trinidad & Tobago
38. Greece	84. Kenya
39. Libya	85. Sierra Leone
40. South-West Africa	86. Yugoslavia
41. Oman	87. Ivory Coast
42. Sweden	88. Guinea
43. China, Republic of	89. German Federal Republic
44. Saudi Arabia	90. Cyprus
45. Turkey	91. El Salvador
46. Iran	92. Malta

Table 3. Continued

93. Yemen (Sana)	106. Lebanon
94. Poland	107. Albania
95. Guatemala	108. Gambia
96. Nauru	109. German Democratic Republic
97. Romania	110. Dahomey
98. Qatar	111. Syria
99. Western Samoa	112. Bahrain
100. Bulgaria	113. Belgium
101. Barbados	114. Togo
102. Congo	115. Zaire
103. Cameroon	116. Iraq
104. Israel	117. Jordan
105. Kuwait	118. Singapore

hydrocarbon fields would have to be most extensive to warrant the expenses of exploration and exploitation. Only 35,800 square nautical miles of continental shelf would be an international zone, nearly all of which would be difficult to exploit, being situated in the Arctic under the circulating pack ice. Under existing conventions, even this meager area runs the risk of being allocated to the coastal States. The 118 coastal States[7] would possess a median allocation of 61,900 square nautical miles of seabed, nearly equal to the median of their national land domains (66,700 square nautical miles).

Likely to oppose the 200-mile concept would be the twenty-nine landlocked States[8] and the ten smaller shelflocked States.[9] Eleven or twelve States, probably those with major maritime interests other than in the seabeds, would need to join forces with this group to form the requisite one-third (assuming 150 participant States and a requirement of a two-thirds vote for passage) required to block the issue. However, if national jurisdiction was limited to 200 meters, the total

7. In the listings South-West Africa has been listed as independent by virtue of G.A. Res. 2145, 21 U.N. GAOR Supp. 1, at 60, U.N. Doc. A/6301 (1966), although South Africa controls South-West Africa's administration. Monaco has not been generally discussed; its seabed area is negligible. Its foreign relations, moreover, are controlled by France. Shelflocked Monaco, however, has retained a great interest in the sea due to the activities of the ruling family and because the International Hydrographic Organization is located in Monaco.

8. The total could be greater if the Soviet Union's constituent republics—the Ukraine and Byelorussia, both U.N. Members—attend a conference. The number is not really relevant except that a majority vote will require many more votes than in 1958 or in 1960.

9. Bahrain, Belgium, German Democratic Republic, Iraq, Jordan, Kuwait, Qatar, Singapore, Togo, and Zaire.

Table 4. Maximum theoretical allocation of seabed with absolute and percentage reductions of other proposals

	200 meters (%)	40 n.m. (%)	Margin (%)	200 n.m. (%)
1. Albania	−2,000 (−55.6)	−300 (−8.3)	3,600*	3,600*
2. Algeria	−36,000 (−90.0)	−17,200 (−43.0)	−26,400 (−66.0)	40,000*
3. Argentina	−251,900 (−52.0)	−406,900 (−84.1)	484,100*	−144.600 (−29.9)
4. Australia	−1,381,700 (−67.6)	−1,634,700 (−80.0)	−597,900 (−29.3)	2,043,300*
5. Bahrain	1,500*	−100 (−6.7)	1,500*	1,500*
6. Barbados	−48,700 (−99.8)	−42,000 (−86.1)	−46,500 (−95.3)	48,800*
7. Bangladesh	−6,400 (−28.6)	−13,200 (−58.9)	−1,600 (−7.1)	22,400*
8. Belgium	800*	800*	800*	800*
9. Brazil	−699,900 (−75.7)	−97,300 (−10.5)	−488,300 (−52.8)	924,000*
10. Bulgaria	−6,000 (−62.5)	−4,700 (−49.0)	9,600*	9,600*
11. Burma	−81,700 (−55.0)	−89,400 (−60.2)	−37,300 (−25.1)	148,600*
12. Cambodia	16,200*	−4,900 (−30.2)	16,200*	16,200*
13. Cameroon	−1,400* (−31.1)	−200 (−4.4)	4,500*	4,500*
14. Canada	−523,500 (−38.2)	−407,000 (−29.7)	−130,000 (−9.5)	1,370,000*
15. Chile	−659,300 (−98.8)	−512,300 (−76.8)	−499,400 (−74.8)	667,300*
16. China, People's Republic of	−50,900 (−18.1)	−142,500 (−50.7)	281,000*	281,000*
17. China, Republic of	−90,900 (−79.5)	−79,500 (−69.5)	−74,500 (−65.1)	114,400*
18. Colombia	−156,100 (−88.7)	−102,300 (−58.2)	−115,000 (−65.4)	175,900*
19. Congo	−4,600 (−63.9)	−4,000 (−55.6)	7,200*	7,200*
20. Costa Rica	−70,900 (−93.9)	−52,000 (−68.9)	−60,400 (−80.0)	75,500*
21. Cuba	−82,500 (−78.0)	−13,900 (−13.1)	−36,900 (−34.9)	105,800*
22. Cyprus	−27,100 (−93.4)	−12,400 (−42.8)	−17,100 (−59.0)	29,000*
23. Dahomey	−7,400 (−93.7)	−5,200 (−65.8)	−5,300 (−67.1)	7,900*

*Largest allocation.

Table 4. Continued

	200 meters (%)	40 n.m. (%)	Margin (%)	200 n.m. (%)
24. Denmark	20,000*	−9,500 (−47.5)	20,000*	20,000*
25. Dominican Republic	−73,100 (−93.2)	−56,900 (−72.6)	−49,500 (−63.1)	78,400*
26. Ecuador	−324,300 (−95.9)	−241,000 (−71.3)	−285,400 (−84.4)	338,000*
27. Egypt	−39,700 (−78.5)	−13,900 (−27.5)	−21,700 (−42.9)	50,600*
28. El Salvador	−21,600 (−80.6)	−21,000 (−78.4)	−14,500 (−54.1)	26,800*
29. Equatorial Guinea	−79,000 (−95.6)	−64,500 (−78.1)	−67,800 (−82.1)	82,600*
30. Ethiopia	−8,200 (−37.1)	−1,100 (−5.0)	22,100*	22,100*
31. Fiji	−330,300 (−99.8)	−256,300 (−77.5)	−329,900 (−99.7)	330,900*
32. Finland	28,200*	−400 (−1.4)	28,200*	28,200*
33. France	−46,400 (−46.6)	−51,000 (−51.3)	−23,700 (−23.8)	99,500*
34. Gabon	−48,900 (−78.5)	−45,700 (−73.4)	−21,400 (−34.3)	62,300*
35. Gambia	−4,000 (−70.2)	−4,500 (−78.9)	−2,400 (−42.1)	5,700*
36. German Democratic Republic	2,800*	2,800*	2,800*	2,800*
37. Germany, Federal Rep. of	11,900*	−1,700 (−14.3)	11,900*	11,900*
38. Ghana	−57,500 (−90.4)	50,900 (−80.0)	−43,500 (−68.4)	63,600*
39. Greece	−140,100 (−95.1)	−61,500 (−41.8)	−65,200 (−44.3)	147,300*
40. Guatemala	−25,300 (−87.5)	−23,100 (−79.9)	−20,700 (−71.6)	28,900*
41. Guinea	−9,500 (−45.9)	−14,100 (−68.1)	−5,400 (−26.1)	20,700*
42. Guiana	−23,400 (−61.6)	−29,000 (−76.3)	−9,700 (−25.5)	38,000*
43. Haiti	−43,700 (−93.4)	−30,000 (−64.1)	−30,800 (−65.8)	46,800*
44. Honduras	−43,000 (−73.4)	−33,000 (−56.3)	−14,700 (−25.1)	58,600*
45. Iceland	−213,800 (−84.6)	−195,300 (−77.3)	−800 (−0.3)	252,800*
46. India	−455,800 (−77.6)	−411,300 (−70.0)	−247,900 (−42.2)	587,600*
47. Indonesia	−767,700 (−48.7)	−546,200 (−34.6)	−347,500 (−22.0)	1,577,300*

Table 4. Continued

	200 meters (%)	40 n.m. (%)	Margin (%)	200 n.m. (%)
48. Iran	−14,200 (−31.3)	−12,200 (−26.9)	45,400*	45,400*
49. Iraq	200*	200*	200*	200*
50. Ireland	−74,200 (−66.9)	−83,400 (−75.2)	−26,800 (−24.2)	110,900*
51. Israel	−5,500 (−80.9)	−2,400 (−35.3)	−1,100 (−16.2)	6,800*
52. Italy	−119,000 (−73.9)	−57,400 (−35.7)	−1,000 (−0.6)	161,000*
53. Ivory Coast	−27,500 (−90.2)	−19,100 (−62.6)	−14,800 (−48.5)	30,500*
54. Jamaica	−75,100 (−86.5)	−59,900 (−69.0)	−27,800 (−32.0)	86,800*
55. Japan	−985,900 (−87.6)	−945,900 (−84.0)	−685,100 (−60.8)	1,126,000*
56. Jordan	200*	200*	200*	200*
57. Kenya	−30,200 (−87.8)	−24,600 (−71.5)	−12,800 (−37.2)	34,400*
58. Korea, North	−24,600 (−65.1)	−16,600 (−43.9)	−17,400 (−46.0)	37,800*
59. Korea, South	−30,300 (−29.8)	−53,300 (−52.5)	−8,300 (−8.2)	101,600*
60. Kuwait	3,500*	−500 (−14.3)	3,500*	3,500*
61. Lebanon	−5,300 (−80.3)	−2,500 (−37.9)	−2,000 (−30.3)	6,600*
62. Liberia	−61,300 (−91.5)	−54,700 (−81.6)	−47,400 (−70.7)	67,000*
63. Libya	−74,200 (−75.3)	−60,500 (−61.4)	−38,500 (−39.0)	98,600*
64. Madagascar	−324,200 (−86.0)	−299,500 (−79.5)	−245,500 (−65.2)	376,800*
65. Malaysia	−29,800 (−21.5)	−72,100 (−52.0)	−13,100 (−9.4)	138,700*
66. Maldives	−276,700 (−98.9)	−215,600 (−77.1)	−275,700 (−98.6)	279,700*
67. Malta	−15,500 (−80.3)	−14,000 (−72.5)	−2,000 (−10.4)	19,300*
68. Mauritania	32,100 (−71.3)	−34,000 (−75.6)	−18,700 (−41.6)	45,000*
69. Mauritius	−318,300 (−92.3)	−314,600 (−91.2)	−195,800 (−56.8)	345,000*
70. Mexico	−702,600 (−84.5)	−595,400 (−71.6)	−488,500 (−58.7)	831,500*
71. Morocco	−63,000 (−77.7)	−50,000 (−61.7)	−39,000 (−48.1)	81,100*
72. Nauru	−125,600 (−99.9)	−120,500 (−95.9)	−125,500 (−99.8)	125,700*

Table 4. Continued

	200 meters (%)	40 n.m. (%)	Margin (%)	200 n.m. (%)
73. Netherlands	24,700*	−13,200 (−53.4)	24,700*	24,700*
74. New Zealand	−1,338,700 (−95.0)	−1,263,100 (−89.6)	−838,400 (−59.5)	1,409,500*
75. Nicaragua	−25,400 (−54.5)	−24,500 (52.6)	−9,200 (−19.7)	46,600*
76. Nigeria	−48,000 (−78.0)	−44,600 (−72.5)	−23,800 (−38.7)	61,500*
77. Norway	−560,500 (−94.9)	−412,500 (−69.9)	−126,800 (−21.5)	590,500*
78. Oman	−146,000 (−89.1)	−127,100 (−77.6)	−119,300 (−72.8)	163,800*
79. Pakistan	−75,900 (−81.7)	−76,500 (−82.3)	−31,500 (−33.9)	92,900*
80. Panama	−72,700 (−81.3)	−52,200 (−58.4)	−49,100 (−54.9)	89,400*
81. Peru	−205,300 (−89.5)	−163,300 (−71.2)	−180,500 (−78.7)	229,400*
82. Philippines	−499,400 (−90.6)	−321,200 (−58.3)	−486,400 (−88.2)	551,400*
83. Poland	8,300*	−500 (−6.0)	8,300*	8,300*
84. Portugal	−506,000 (−97.8)	−456,800 (−88.3)	−472,600 (−91.3)	517,400*
85. Qatar	7,000*	−400 (−5.7)	7,000*	7,000*
86. Romania	−2,200 (−23.7)	−4,800 (−51.6)	9,300*	9,300*
87. Saudi Arabia	−32,200 (−59.3)	−10,500 (−19.3)	−300 (−0.6)	54,300*
88. Senegal	−50,800 (−84.7)	−49,200 (−82.0)	−36,300 (−60.5)	60,000*
89. Sierra Leone	−37,700 (−83.0)	−36,800 (−81.1)	−30,700 (−67.6)	45,400*
90. Singapore	100*	100*	100*	100*
91. Somalia	−210,600 (−92.2)	−164,100 (−71.9)	122,100 (−53.5)	228,300*
92. South Africa	−254,700 (−85.9)	−239,500 (−80.8)	−112,800 (−38.0)	296,500*
93. South-West Africa	−126,900 (−87.0)	−118,000 (−80.9)	−73,200 (−50.2)	145,900*
94. Soviet Union	−945,200 (−72.2)	−452,300 (−34.5)	−573,600 (−43.8)	1,309,500*
95. Spain	−305,900 (−86.0)	−229,600 (−64.6)	−197,400 (−55.5)	355,600*
96. Sri Lanka (Ceylon)	−143,100 (−94.8)	−105,600 (−70.0)	−124,000 (−82.2)	150,900*

Table 4. Continued

	200 meters (%)	40 n.m. (%)	Margin (%)	200 n.m. (%)
97. Sudan	-20,200 (-75.7)	-8,400 (-31.5)	-200 (-0.7)	26,700*
98. Sweden	-100 (-0.2)	-1,200 (-2.6)	45,300*	45,300*
99. Syria	-1,900 (-63.3)	-100 (-3.3)	3,000*	3,000*
100. Tanzania	-53,100 (-81.6)	-41,500 (-63.7)	-23,500 (-36.1)	65,100*
101. Thailand	-19,600 (-20.7)	-37,600 (-39.7)	94,700*	94,700*
102. Togo	300*	300*	300*	300*
103. Tonga	-169,600 (-97.6)	-131,400 (-75.6)	-161,800 (-93.1)	173,800*
104. Trinidad & Tobago	-13,900 (-62.1)	-11.400 (-50.9)	-5,100 (-22.8)	22,400*
105. Tunisia	-10,200 (-40.8)	-2,300 (-9.2)	25,000*	25,000*
106. Turkey	-54,300 (-78.7)	-25,200 (-36.5)	69,000*	69,000*
107. United Arab Emirates	17,300*	17,300*	17,300*	17,300*
108. United Kingdom	-138,300 (-49.1)	-176,900 (-62.8)	281,800*	-7,000 (-2.5)
109. United States	-1,676,600 (-75.5)	-1,490,100 (-67.1)	-1,359,400 (-61.2)	2,222,000*
110. Uruguay	-18,300 (-46.2)	-28,000 (-70.7)	39,600*	-4,800 (-12.1)
111. Venezuela	-80,400 (-75.8)	-49,800 (-46.9)	-36,400 (-34.3)	106,100*
112. Vietnam, North	22,200*	-5,000 (-22.5)	22,200*	22,200*
113. Vietnam, South	-92,800 (-49.3)	-138,900 (-73.7)	-37,000 (-19.6)	188,400*
114. Western Samoa	-26,800 (-95.7)	-17,200 (-61.4)	-25,500 (-91.1)	28,000*
115. Yemen (Aden)	-145,400 (-90.6)	-116,400 (-72.5)	-70,400 (-43.9)	160,500*
116. Yemen (Sana)	-1,700 (-17.2)	-900 (-9.1)	9,900*	9,900*
117. Yugoslavia	-4,600 (-30.1)	-400 (-2.6)	15,300*	15,300*
118. Zaire	300*	300*	300*	300*

national area of seabed would be reduced to 8,031,000 square nautical miles while the median would decrease by more than three-quarters to 13,600 square nautical miles.[10] Thus, 29,714,000 square nautical miles would be added to the international zone. Depending on the regime, the rent or royalties from this potentially more fruitful zone would be paid to a development fund for the less-developed States. Of course, an important consideration would be to keep operating machinery costs to a minimum in order not to reduce drastically the economic benefits of the proposal. If the machinery were too complicated, and thus too dear, the coastal States might deem it wiser to choose national jurisdiction over the widest possible area.

An examination of the relative effects of the various seabed proposals reveals interesting correlations. For example, States with longest coastlines, which tend to be the largest and best-developed States, gain the most from expansive seabed proposals. It stands to reason that a State with a 1,000-mile coastline will obtain a greater portion of the adjacent seabed than one with a 100-mile coastline. Two other factors influence the allocation: location on a semienclosed sea, which tends to reduce the absolute value of a long coast in case of a mileage-determined seabed boundary, and insularity, which greatly expands the effect of the coast-length factor under a mileage criterion but generally decreases it on a depth criterion. Thus, while India's coast measures twice that of France, the areal allocations rate is more than 3 to 1. The allocations rate is greater because of France's extensive adjacency to the United Kingdom, Spain, and Italy and India's extensive insularity in the Andaman and Laccadive Islands and its bordering only on Sri Lanka (Ceylon) and the Maldives for short distances. Another example of this phenomenon is Peru and Saudi Arabia, which have nearly identical coastal lengths. Peru, however, is situated on the open ocean, while Saudi Arabia borders two semienclosed seas. On a 200–meter seabed limit, Peru ranks 35th and Saudi Arabia 39th of the world's coastal States. On the other hand, under a 200-mile limit, Peru rises to 25th while Saudi Arabia falls to 63rd. Further illustrating this disparity in ranking are Lebanon and Fiji,

---

10. *See* note 3 *supra.* The seven power working paper proposes the system of 40 n.m. or 200 meters as the national limit. Certain States, of course, would have no problem of selection. A 40-mile limit would be everywhere landward or seaward of the 200-meter limit. Most, however, would have a difficult choice. The United States, for example, would be forced to choose losing shelf area in the Gulf of Mexico and the East Coast in order to gain area on the West Coast and Hawaii or vice versa. Since the current Convention on the Continental Shelf guarantees a national shelf limit of 200 meters, we feel that the proposal, to be considered seriously, must be expressed as 40 n.m. *and* 200 meters, whichever limit may be farther seaward.

which have nearly the same coastal lengths. A 200-meter limit ranks them 106th and 110th, respectively, while a 200-mile limit puts them in 105th and 19th place. Fiji's open-ocean insularity negates all other factors on a distance seabed limit.

As a generalization, countries situated on semienclosed seas can be expected to oppose distance criteria and favor a depth criterion. In contrast, island nations can be expected to oppose a depth criterion and favor those based upon distance. Some nations, of course, are often confronted with both of these factors. The United States is situated on the Gulf of Mexico, a semienclosed sea, but also is extended to open islands in the Aleutian and Hawaiian Islands. This analysis applies only to national territory, not to dependencies administratively separate from the State. The analysis reveals that eleven countries, in rank order, fare very well under all criteria: the Soviet Union, Indonesia, Australia, the United States, Canada, Japan, Mexico, Brazil, India, New Zealand, and Norway. These States, the top twenty-two according to coastal lengths, rank fifteenth or above in all areal seabed allocations. The other long-coastline States[11] rank well in three of the four criteria but suffer in one, usually in the 200–nautical mile category due to their location on semienclosed seas. It also becomes obvious that nineteen States really cannot expect to gain much seabed from any of the proposals.[12] These States rank in the lowest twenty-eight positions in all categories. Primarily, they have short coastlines, are entrapped by convex coastal configurations, or are on semienclosed seas.[13] Singapore, for example, occupies last place in all categories.

Thus, four categories of States immediately appear: (1) twenty-nine landlocked States;[14] (2) nineteen shelflocked States;[15] (3) eleven States

11. Philippines, People's Republic of China, Chile, United Kingdom, Italy, Spain, Argentina, Malaysia, Madagascar, and Turkey.
12. Albania, Bahrain, Belgium, Bulgaria, Cameroon, Congo, Dahomey, Gambia, German Democratic Republic, Iraq, Israel, Jordan, Kuwait, Lebanon, Qatar, Singapore, Syria, Togo, and Zaire.
13. One of the difficulties in relating choices to area is illustrated by Kuwait. Ranking between 94th (200 meters) and 109th (200 miles) with an areal allocation of 3,500 sq. n.m. under all parameters, Kuwait virtually floats on oil. Moreover, its rank is commensurate with its land area (102d) and coastal length (97th).
14. Afghanistan, Andorra, Austria, Bhutan, Bolivia, Botswana, Burundi, Central African Republic, Chad, Czechoslovakia, Hungary, Laos, Lesotho, Liechtenstein, Luxembourg, Malawi, Mali, Mongolia, Niger, Nepal, Paraguay, Rwanda, San Marino, Swaziland, Switzerland, Uganda, Upper Volta, Vatican City, Zambia.
15. Bahrain, Belgium, Cambodia, Denmark, Finland, German Democratic Republic, German Federal Republic, Iraq, Jordan, Kuwait, the Netherlands, Poland, Qatar, Singapore, Sweden (virtually), Togo, United Arab Emirates, North Vietnam, Zaire.

that fare well and are relatively unaffected by differences in seabed proposals; and (4) nineteen States that, relatively speaking, cannot be helped. The four groups, which admittedly form a disparate coalition, total sixty-eight States.[16]

Another grouping, though much less definite and distinctive, may be delineated. These States, situated in the middle between those relatively unhurt and those relatively unhelped by the various seabed proposals, vary only slightly in relatively rank order and may be classified as relatively unaffected. Morocco typifies these thirty-nine States.[17] It ranks 40th in coastal length, 49th according to the margin, and 44th by a 200-meter limit; by the criteria of 40 and 200 nautical miles, it ranks 50th and 52nd, respectively. No matter which seabed boundary criterion is selected, Morocco's national seabed allocation position remains relatively constant and is commensurate with its geographical size and coastal length. If a seabed boundary selection is to be made on a relative basis, the former group of sixty-eight and this latter group of thirty-nine States will have to form the core of a consensus.

A third method of analysis is based upon the negative relative effects of the various proposals. Stated simply, one may ask which States have marked changes in relative position under the specific proposals. Only eight States[18] suffer a marked reduction in relative standing according to a 200-meter–40-nautical-mile seabed proposal. Significantly, five are island States or contain important island elements within their boundaries.[19] In contrast, twenty-eight States[20] experience considerable relative decrease in rank based on a 200-nautical mile criterion for the limit of national jurisdiction; of that

16. Ten States occur on two of the lists.
17. Chile, Cuba, Cyprus, Dominican Republic, Egypt, El Salvador, France, Gabon, Greece, Guiana, Haiti, Honduras, Iceland, Jamaica, Kenya, North Korea, Libya, Madagascar, Malta, Mauritania, Morocco, Nicaragua, Nigeria, Oman, Pakistan, Panama, Senegal, Sierra Leone, Somalia, South Africa, South-West Africa, Spain, Tanzania, Trinidad and Tobago, Turkey, Venezuela, Yemen (Aden), Yemen (Sana), Yugoslavia.
18. Barbados, Ghana, Guatemala, Liberia, Mauritius, Nauru, Portugal, Tonga.
19. Note that a country may have deficiencies in one or two categories and hence be found in two of the three listings. In fact, nearly all States that suffer relatively from the 200-meter–40-n.m. criteria also experience relative decrease with the margin limit. Only Mauritius is an exception, and this may relate to the method of allocation, which required considerable subjective judgment.
20. Argentina, Bangladesh, Burma, Cambodia, People's Republic of China, Denmark, Ethiopia, Finland, Federal Republic of Germany, Guinea, Iran, Iceland, Italy, South Korea, Malaysia, the Netherlands, Poland, Romania, Saudi Arabia, Sweden, Thailand, Tunisia, United Arab Emirates, United Kingdom, Uruguay, North Vietnam, South Vietnam, Yugoslavia.

group twenty-four States are situated wholly, or to a major degree, on semienclosed seas. Only Argentina, Bangladesh, Guinea, and Uruguay—all open-ocean States—seem to be affected negatively by extensive shelf areas, a factor that operates to restrict additional areas under a 200–nautical mile limit. They gain so much seabed under a shelf definition that the 200–nautical mile extension causes a significant decrease in relation to the increased gain by other, less well-endowed states. For example, Argentina's percentage of the national seabed area decreases from 3.75 percent with a 200-meter boundary to 1.38 percent with a 200–nautical mile limit.

Finally, with the margin limitation, thirty States[21] suffer a marked decrease in rank. As noted, these include the seven shelf-poor States but also twelve shelflocked or semishelflocked States (Bangladesh, Cambodia, Denmark, Finland, the German Federal Republic, Iran, the Netherlands, Poland, Romania, Sweden, North Vietnam, and Guinea). These States represent the ones least likely to favor the United States seabed proposal.

From a positive viewpoint, which proposal favors each State? Surprisingly, a 200-meter–40-nautical-mile criterion is relatively the most favorable to sixty-nine coastal States and twenty-nine land-locked States and thus gives relative advantage to a total of ninety-eight coastal and landlocked States.[22] As Maps 7–9 illustrate, these countries are primarily situated on semienclosed seas or along the Atlantic margins of the great shelf areas. In contrast, the 200–nautical mile boundary is favorable to only 28 States,[23] which are primarily island States or are situated on the western shores of the continents, where shelves tend to be narrower. The margin criterion also favors twenty-eight States that are scattered randomly in no set geographic pattern. Significantly, however, only two are situated in the Americas.

Measurement of the absolute areal effect of various seabed proposals on coastal States is determined by calculating the loss a State incurs from consideration of a seabed proposal in comparison to the maximum seabed areal allocation. In calculating the absolute areal change, the negative result of a seabed proposal probably would

21. Bangladesh, Barbados, Cambodia, Republic of China, Colombia, Denmark, Ecuador, Equatorial Guinea, Fiji, Finland, Federal Republic of Germany, Ghana, Guatemala, Guinea, Iran, Liberia, the Maldives, Nauru, the Netherlands, Peru, Philippines, Poland, Portugal, Romania, Sri Lanka (Ceylon), Sweden, Tonga, United Arab Emirates, North Vietnam, Western Samoa.
22. *See* Maps 7–9.
23. Several States have the same rank order under two categories and hence are listed twice; the total, as a result, exceeds 118. Singapore, ranking 118th in all categories, is not listed in any.

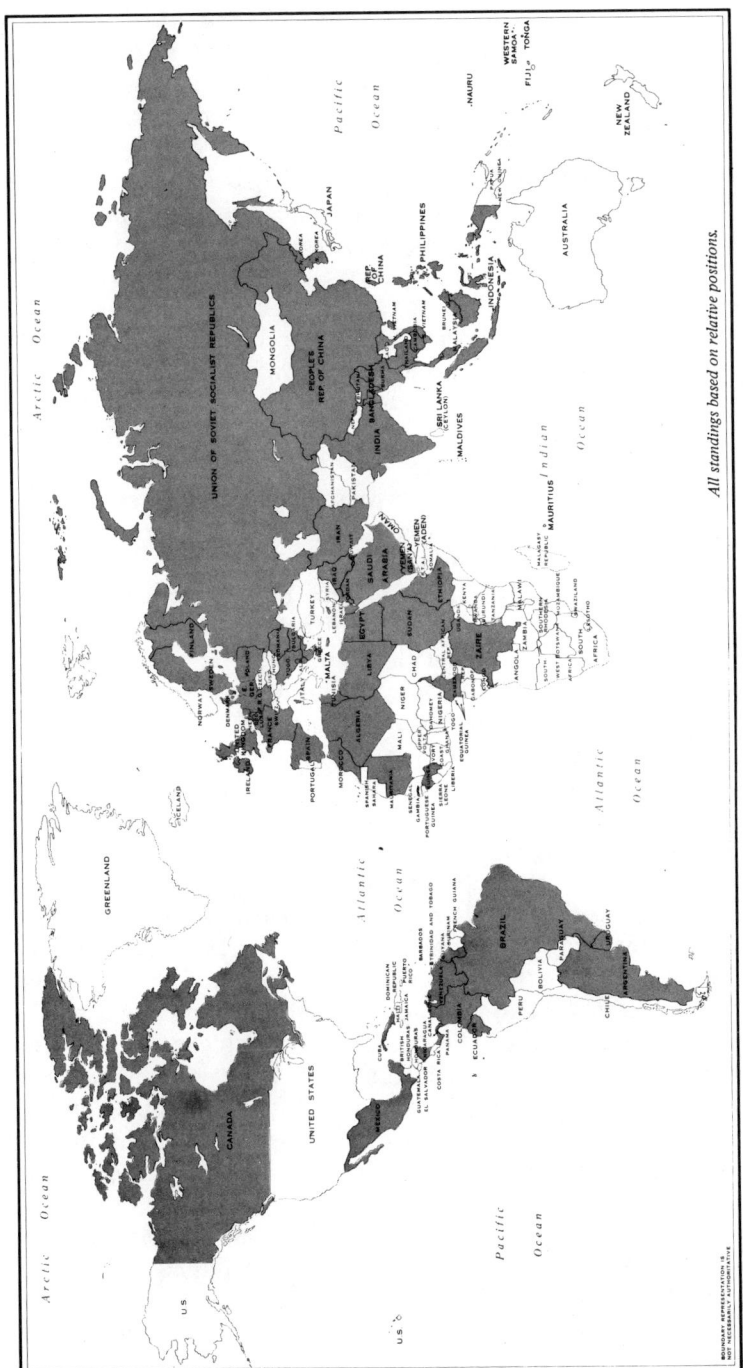

7. Coastal States benefited most from the 200 meter / 40 nautical mile and 200 meter criterion

All standings based on relative positions.

provide more of an impact on a State's perception than the positive absolute effects of an alternative proposal. That is to say, many absolute changes on the positive side are zero or negligible, implying that the coastal State has achieved its maximum seabed allocation under one or more proposals.

Table 4 showing the maximum theoretical allocation of seabed with absolute and percentage reductions of other proposals, immediately makes it apparent that the 200–nautical mile proposal provides the largest areal allocation to the coastal States with the exceptions of Argentina, the United Kingdom, and Uruguay. Therefore, if a nationalistic point of view prevails in the seabed negotiations, the 200–nautical mile seabed proposal will prevail with little opposition, given the premise of maximum areal seabed allocations to the coastal States and the minimum to the common heritage of mankind or the internationally controlled regime.

If the seabed negotiations are to preserve the maximum area of the seabed for mankind or the international regime, an alternative seabed boundary must replace the 200–nautical mile proposal. The proposal of the Committee of Seven gives to each State the maximum areal allocation obtainable under either a 40–nautical mile or 200–meter seabed boundary and the remainder to the international regime. This proposal would provide the maximum area to the international regime. Table 4 also reveals that no State achieves a maximum areal allocation under the 40–nautical-mile—200-meter criteria that is not obtainable under another proposal. Therefore, acceptance of the proposal is essentially a reduction of seabed area of the coastal States with an allocation to the international regime.

How, then, will States examine and evaluate their seabed options on the basis of the area data? If an internationalist climate prevails, the value of relative rank order reflects the best potential selection for a national option within the international framework. Thus, it would appear that a 200-meter national limit for seabed jurisdiction is most favorable to most nations. The twenty-nine landlocked States would gain the greatest economic advantage from the minimum national seabed proposal, that is, the 200–meter—40–nautical-mile boundary. As stated, sixty-nine other States rank highest or higher in the same category, and a total of ninety-nine nations would be best served, relatively, under this concept.

However, many of the States that fall into this grouping have publicly disavowed any interest in the limit and stated their intent to support a wider national boundary on the seabed. Unless Argentina, Brazil, Ecuador, and Uruguay, among others, can be convinced of the

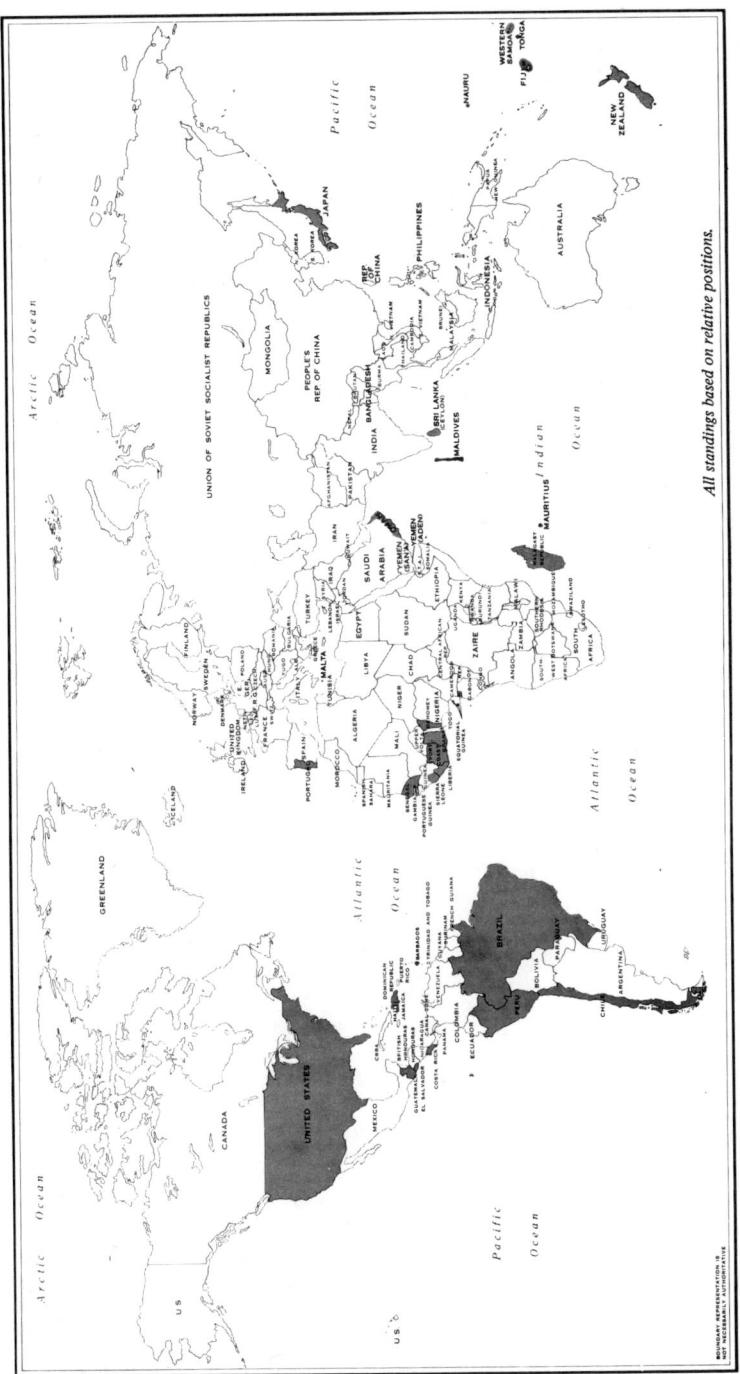

8. Coastal States benefited most from the 200 nautical mile criterion

All standings based on relative positions.

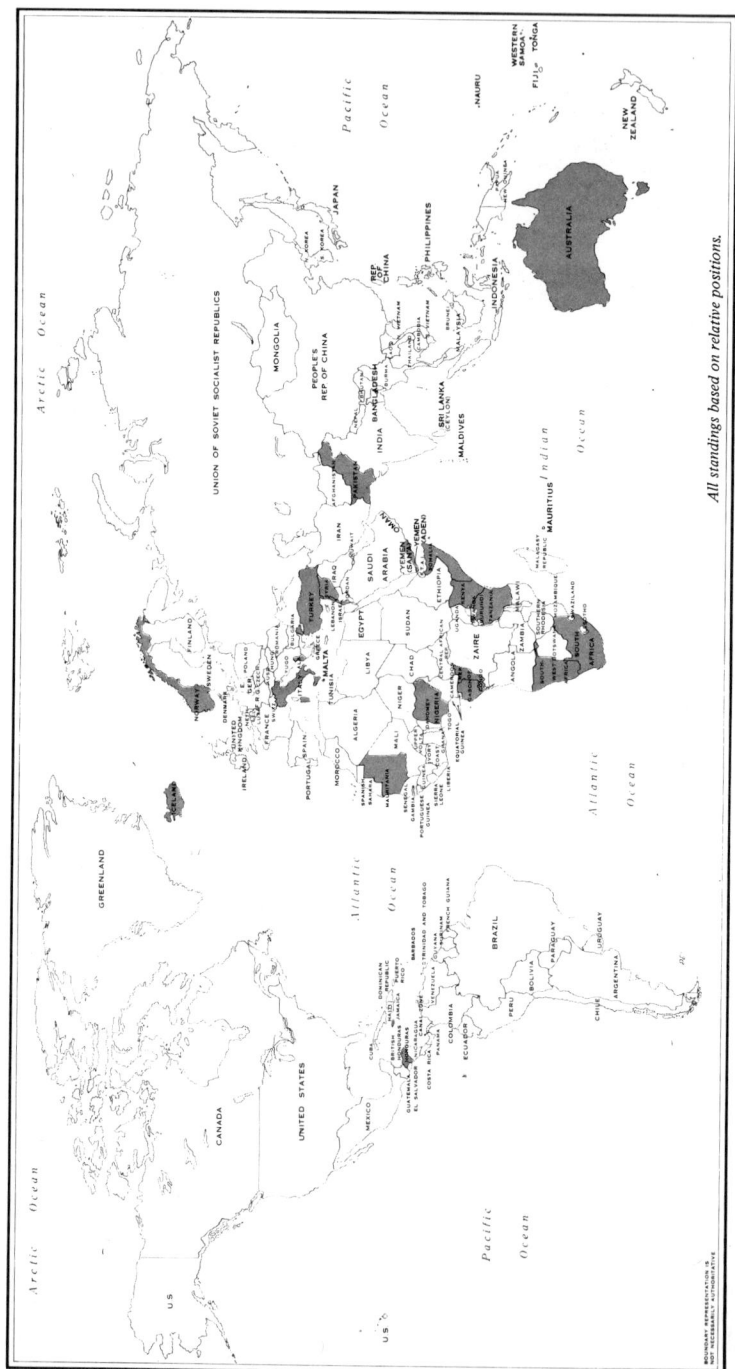

9. Coastal States benefited most from the continental margin criterion

All standings based on relative positions.

merits of a 200-meter limit, a compromise will have to be negotiated to maintain the dominant position of the group of ninety-nine.

Public statements and the official debates in Geneva and New York, however, indicate that the common heritage concept has come to equate practically with a maximum national claim to the seabed. This fact appears to be true at least for the developing States, which depend on primary production for their economic well-being. If this condition does prevail, there is no question about the ultimate national options for the boundary between national and international jurisdiction on the seabeds. Only three coastal States—Argentina, Uruguay, and the United Kingdom—would receive a greater seabed allocation under a margin boundary than under a 200–nautical mile limit for national jurisdiction. One hundred fifteen coastal nations would receive the maximum national seabed territory under a 200-mile zone.

From the negative point of view, the twenty-nine landlocked States and the nineteen (twenty with Monaco) shelflocked States have nothing to gain and, perhaps, much to lose from any limit wider than 200 meters–40 nautical miles. Notably, these two groups combined should be able to block any proposal detrimental to their interests. Regional and other ideological pressures, however, may tend to weaken the political effects of this grouping, but eight semishelflocked States may form potential allies of this group to counteract these pressures.[24] Thus, an analysis of the data indicates both a potential for a strong minority, which can block the adoption of a proposal, and a majority for a 200–nautical mile seabed boundary. Neither of the remaining proposals would appear to generate sufficient appeal for approval on an absolute basis, although a 200-meter–40-nautical-mile limit has potential strength. Consequently, absent other overriding pressures, diplomatic negotiations may favor either a 200–nautical mile exclusive seabed boundary if a nationalistic atmosphere prevails or a combination of 200 meters and 40 nautical miles for the limit of national jurisdiction and a 200–nautical mile preference zone, similar to the United States intermediate trusteeship zone, should there be a compromise of national and international interests. The latter option, however, would have to overcome strong opposition based on emotional or political factors to attain a majority of votes in a future conference on the law of the sea. Thus, as in other areas of international law, the importance of the political climate should not be underestimated.

24. Bulgaria, Cameroon, People's Republic of China, Romania, Saudi Arabia, Sudan, Yemen (Sana), and Yugoslavia.

Contemporary State Practice

V

The USSR and the Limits to National Jurisdiction over the Sea, 1970–72

William E. Butler

It is only a slight exaggeration to say that East-West differences dominated the 1958 and 1960 Geneva Conferences on the Law of the Sea; much was accomplished despite profound political and ideological divergencies. By the same token, the insoluble issues, particularly the breadth of the territorial sea, foundered precisely because the major powers believed that the distance between their positions (literally, nine miles at the 1958 Conference, six miles in 1960) was unbridgeable without undue advantage accruing to the other side.

Rarely in international law have premises so firmly held been dashed so quickly. A mere six years following the abortive 1960 Geneva Conference, the pillar of support for the three-mile rule, the United States, had created a twelve-mile fishing zone—partly to protect coastal fish stocks from Soviet flotillas. By 1969 the United States had officially expressed support for a twelve-mile territorial sea provided that appropriate guarantees of free passage through international straits could be secured in an international agreement. For its part, the Soviet Union has retracted its view that States ought to fix the breadth of their territorial sea in accordance with their own special interests. Soviet jurists now believe that twelve miles is the absolute maximum territorial sea admissible under international law and that free passage through straits must be ensured. These are merely two issues of many on which the major powers now find themselves in basic accord. Paradoxically, however, it is unlikely that the combined efforts of the most powerful maritime powers will produce an international treaty wholly satisfactory to them, so radically have their interests in the seas (and their capacity to influence them) altered during the past decade.

In 1970 the present writer was so bold as to suggest: "Broadly speaking, the development of Soviet attitudes toward the international law of the sea falls into three distinct periods; we may be on the verge of yet a fourth."[1] The fourth period was in fact well underway. What

---

1. W. BUTLER, THE SOVIET UNION AND THE LAW OF THE SEA 198 (1971).

follows is an attempt to appraise developments in Soviet attitudes from 1970 to 1972, updating my earlier study in the light of new materials and sharpening somewhat the periodization postulated earlier.

The Territorial Sea

Following the failure of the 1960 Geneva Conference on the Law of the Sea to reach agreement on the breadth of the territorial sea, the Soviet Government unilaterally acted to realize a long-cherished objective: on August 5, 1960, a new Statute on the Protection of the State Border of the USSR formally established a twelve-mile territorial sea.[2] This action was in accord with the long-standing Soviet position that such a limit was justified under prevailing international law and, moreover, that States were entitled to fix the breadth of their territorial sea in conformity with their economic and security interests. The Statute stipulated that territorial waters were to be measured from the line of lowest ebbtide or the seaward line of internal sea waters.

Since 1960 Soviet policy with respect to territorial waters has changed in two respects. In order to take full advantage of the methods admissible under the 1958 Geneva Convention on the Territorial Sea and Contiguous Zone for delimiting territorial waters,[3] the 1960 Statute was amended in 1971[4] to permit the use of straight base lines in locales where the coastline is deeply indented and cut into or where there is a fringe of islands along the coast in its immediate vicinity. Although neither the amendment nor the relevant provision of the 1958 Convention stipulates the length of base lines, there is support in Soviet doctrinal literature for limiting them to twenty-four nautical miles.[5]

The twelve-mile limit is now firmly entrenched in Soviet legal doctrine as the optimal breadth for the territorial sea. Citing the view of the International Law Commission that the practice of States in fixing the breadth of the territorial sea is not uniform and that international law does not permit an extension of the territorial sea beyond twelve

2. [1960] 34 Ved. Verkh. Sov. S.S.S.R. Item 324 (Supreme Soviet USSR) art. 3, 6 Sov. Stat. & Dec. No. 1, at 45 (1969).

3. *Done,* Apr. 29, 1958, [1964] 2 U.S.T. 1606, T.I.A.S. No. 5639, 516 U.N.T.S. 205.

4. For the text of Article 3 as amended, see Butler, *New Soviet Legislation on Straight Base Lines,* 20 Int'l & Comp. L. Q. 750–52 (1971).

5. For a review of Soviet doctrinal writings, see *id.* at 751.

miles, L. Speranskaia[6] has pointed to the influence of this view on the maximum extent of a contiguous zone as laid down by Article 24(2) of the 1958 Convention: "If the contiguous zone must not exceed the twelve-mile limit, naturally the territorial sea also cannot exceed it if its breadth does not coincide with this limit."[7]

The actions of Latin American coastal States in creating a 200-mile territorial sea are deplored by Soviet jurists as an understandable but misguided effort to protect fishery resources. They cannot be regarded as "lawful, since they violate generally recognized norms and principles of international law and inflict damage on the rights and interests of other countries."[8] The Soviet stake in this issue is a material one. Khlestov, legal adviser to the USSR Ministry of Foreign Affairs, reminded the United Nations Seabed Committee that if all States adopted a 200-mile limit, as much as 50 percent of the world's oceans would become territorial waters, including some of the most productive fishing grounds now exploited by Soviet fleets.[9] He favored a maximum twelve-mile limit, with a fishery zone also up to twelve miles. Speranskaia seemed somewhat more flexible in concluding that "the question of the limit of territorial waters can no more be decided by each State for itself individually; it must be decided at an international conference, unconditionally, taking into account the geographic peculiarities of various States."[10]

Innocent passage. Debate continues in Soviet legal writing over the precise juridical nature of the limitation of State sovereignty over territorial waters created by the right of innocent passage. Speranskaia is disposed to regard it as a series of obligations assumed by States in respect to foreign vessels, thereby sanctioning the creation of a special regime for one segment of their territory.[11] Of particular interest in connection with innocent passage is Speranskaia's recollection of a Soviet diplomatic Note to Ceylon of February 22, 1964, endorsing a Ceylonese decision to bar from its ports and territorial waters foreign naval vessels carrying nuclear weapons or equipped to conduct nu-

6. Speranskaia in Okean, tekhnika, pravo 54–55 (M. Lazarev & L. Speranskaia eds. 1972) [hereinafter cited as Okean]. Speranskaia reports that as of June 1, 1972 some 55 States had a 12-mile limit and 44, less than 12 miles. Only 14 littoral States of the 113 coastal States on this planet have, according to her information, exceeded the international legal norm. *Id.* at 55.

7. *Id.* at 54–55.

8. *Id.* at 56.

9. U.N. Doc. A/AC. 138/SR. 56, at 150 (1971).

10. Speranskaia, *supra* note 6, at 60.

11. *Id.* at 69.

clear warfare. In the Note, the Soviet Government expressed the hope that other nuclear powers would adhere to this position.[12]

In the Soviet view, there is no generally recognized principle of innocent passage for foreign warships in international law. Article 23 of the 1958 Convention, Speranskaia argues rather obscurely, "is binding only on States that have not made a reservation thereto."[13] Nor, she adds, is there a corresponding generally accepted norm of customary international law equivalent to Article 23. After reviewing the legislation of several States on this matter, she draws a conclusion unprecedented in Soviet international legal writing on the subject: "Thus, practice formed over a long period of time demonstrates that unless a coastal State has provided otherwise in one of its legislative acts, foreign warships do not have a right of passage through its territorial waters."[14]

Scientific research. Soviet interests in an appropriate legal regime for scientific research on the high seas are substantial. But proposals such as those advanced by the Stratton Commission, to permit one State to conduct certain scientific research within the territorial sea of another, though described in Soviet legal writing, are looked upon with a high degree of skepticism.[15]

Contiguous zones. As the Soviet Government has come to regard a twelve-mile territorial sea as the maximum admissible limit under international law, Soviet jurists have taken an increasingly jaundiced view of contiguous or special zones exceeding that limit. Writing in 1962, Romanov defined a contiguous zone as "an expanse of the high seas of a specific breadth adjacent to the seaward boundary of the territorial sea and in which the coastal State exercises control in specially provided areas and to a limited extent with regard to its own and to foreign vessels."[16] This definition has been criticized for its failure to

---

12. The Note is published in SBORNIK REGIONAL'NYKH SOGLASHENII I ZAKO-NODATEL'NYKH AKTOV ZARUBEZHNYKH GOSUDARSTV PO VOPROSAM MOREPLA-VANIIA 257–58 (1969).
13. Speranskaia, *supra* note 6, at 70. Speranskaia regards the Law of Feb. 14 1970, adopted by the People's Republic of Yemen, which requires *all* foreign vessels to obtain authorization before entering Yemeni territorial waters, as contrary to international law.
14. *Id.* at 72.
15. *See* COMM'N ON MARINE SCIENCE, ENGINEERING AND RESOURCES, OUR NATION AND THE SEA 203–04 (1969).
16. Romanov in OCHERKI MEZHDUNARODNOGO MORSKOGO PRAVA 128 (V. Koretskii & G. Tunkin eds. 1962).

stipulate a maximum breadth for such zones and to mention the objects for which such zones may be created.[17]

Of the four types of contiguous zones specified in the 1958 Convention on the Territorial Sea (customs, fiscal, immigration, and sanitary), Serkov is especially critical of the 100-mile zone created by the United States in 1935 for customs and fiscal purposes. Then he turns to other special maritime zones not provided for by norms of international law: zones of criminal and civil jurisdiction, security zones, neutrality zones, fortified zones, marine control zones, inspection zones, fishing zones, conservation zones, and areas temporarily closed to navigation. Only the last is expressly endorsed, though presumably the others would not be objectionable so long as the twelve-mile limit were observed.[18] In the absence of a uniform breadth for the territorial sea, Serkov believes a maximum limit of twelve miles for special zones would be optimal. He regrets that international law has not consolidated the rights of coastal States to establish fishery and security zones, a task that requires the most expeditious international settlement. Special noncontiguous zones fixed by international treaty on the model of the 1954 London Convention for the Prevention of Pollution of the Sea by Oil are also, in his view, likely to become more commonplace;[19] he has in mind particularly the establishment of special zones around stations used to collect ocean data.[20]

Internal Sea Waters

A precise definition of historic bays or historic waters has been as elusive in Soviet international legal theory as elsewhere. Until com-

17. *See* Serkov, *Pravovoi rezhim spetsial'nykh morskikh zon,* in AKTUAL'NYE PROBLEMY SOVREMENNOGO MEZHDUNARODNOGO MORSKOGO PRAVA 19 (M. Lazarev *et al.* eds. 1972) [hereinafter cited as AKTUAL'NYE PROBLEMY]. Serkov endorses Kolodkin's definition: "an expanse of high seas situate directly beyond the seaward boundary of territorial waters and having a breadth of not more than 12 miles computed from the same base line as the breadth of territorial waters in which the coastal state exercises control in individual specially stipulated areas for the purpose of not permitting violations of the legal order within its territory and of punishing such violations." Kolodkin in 3 KURS MEZHDUNARODNOGO PRAVA V SHESTI TOMAKH 226 (V. Chkhikvadze *et al.* eds. 1967). Serkov would amend Kolodkin's formulation by making it absolutely clear that 12 miles is the absolute maximum for such zones and that those exceeding the said limit are in breach of the 1958 Convention on the Territorial Sea. Serkov, *supra,* at 19.
18. Serkov, *supra* note 17, at 21–26. He expressly distinguishes between zones around continental shelf installations and zones established pursuant to the 1954 London Convention for the Prevention of Pollution of the Sea by Oil.
19. *Done* May 12, 1954, [1961] 3. U.S.T. 2989, T. I. A. S. 4900, 327 U.N.T.S. 3.
20. *Id.* at 33–35.

paratively recently, Soviet doctrinal literature was replete with examples of historic waters washing the coasts of the Soviet Union. In various books and articles published since the end of World War II, the Gulf of Riga, the Sea of Azov, and the Kara, Laptev, East Siberian and Chukchi Seas have been described as historic waters. The White Sea has had the distinction of being classified variously as a closed sea, an internal Russian sea and a historic bay.[21]

Soviet practice does not throw a great deal of light on the issue. On July 21, 1957, the Council of Ministers of the USSR adopted a decree formally designating Peter the Great Bay to be a part of Soviet internal waters and, except for shipping lanes to Nakhodka, closed to foreign vessels and aircraft.[22] Never before had the bay been mentioned in Soviet legal writing as a historic water.

The 1958 Convention on the Territorial Sea and the Contiguous Zone does not define historic bays.[23] Nor does the 1960 Statute on the Protection of the State Border of the USSR define or expressly designate historic waters, although it contains a reference to "bays, inlets, coves, and estuaries, seas, and straits, historically belonging to the USSR."[24]

Nechaev, a Soviet jurist, has suggested recently that a State must have exercised authority over historic waters for an extended period of time and that the overwhelming majority of States must not have objected to the coastal State declaration that such waters are historic.[25] In the case of a historic bay, the body of water must additionally possess the geographic configuration of a bay and, above all, have economic and defense significance for the coastal State. Doctrinal literature, he points out, generally does not insist that the coastal State submit its claim to a historic bay on condition that it will be recognized by other States.

The omission of a broad and ritualistic listing of historic waters in a Soviet maritime law textbook published in 1969 suggests that Soviet doctrine might be undergoing reconsideration.[26] The White Sea was mentioned as a sea of the bay type, and Peter the Great Bay as an

---

21. *See* W. BUTLER, *supra* note 1, at 107–08. For a discussion of the historic bay doctrine in international law, see Note, *International Law and the Delimitation of Bays,* 49 N.C.L. REV. 943 (1971).

22. Decree of the Council of Ministers of the USSR Concerning Peter the Great Bay, *reported in* Izvestia, July 21, 1957, at 1, cols. 1–2, 6 SOV. STAT. & DEC. No. 2, at 209 (1969–70).

23. *Done* Apr. 29, 1958, [1964] 2 U.S.T. 1606, T.I.A.S. No. 5639, 516 U.N.T.S. 205.

24. *See* text at note, 2 *supra.*

25. Nechaev in OKEAN, *supra* note 6, at 46–48.

26. A. VOLKOV, MORSKOE PRAVO 119 (1969), *translated as* MARITIME LAW (E. Gordon trans. 1970).

example of historic waters. Nechaev refers to the same two bodies of water as traditionally internal and adds a third: the Chesha Bay in the Barents Sea, which "has, just as Peter the Great Bay, integral economic links with the territory enclosing it."[27] So far as the public record is concerned, however, the claim to Chesha Bay is doctrinal in character.

Nechaev also designates the Laptev and Sannikov Straits, which link the Laptev and East Siberian Seas, as historic straits. This claim is believed to be based on an exchange of diplomatic correspondence in 1965 between the Soviet Union and the United States arising out of a contemplated oceanographic research voyage by an American vessel through the Northeast Passage.[28]

Barabolia describes historic straits as

being situated apart from basic routes of international navigation and for a long period of time used only by one coastal State or leading to historic bays and seas. A peculiarity of historic straits consists in the fact that usually a coastal State expends enormous resources to exploit such straits, which go primarily to study the strait, create navigational equipment and signal systems, remove dangers, establish deep channels, and so forth.

Such straits have important economic and defense significance for the coastal State.

The regime of navigation in such straits is completely regulated by coastal State legislation.

Merchant vessels in these straits proceed along previously stipulated routes and pilotage may be prescribed therein, since these straits in fact lead to shores and ports of that State to which they appertain. Warships of other States may traverse historic straits only after obtaining the authorization of the coastal State.[29]

An essential element to sustain a claim to historic straits, in Barabolia's view, is the effective exercise by the coastal State of its sovereign rights for a prolonged period during which others have not used the straits.

The Closed Sea

Although the closed, or regional, sea doctrine has not been accorded the prominence in recent Soviet legal writings on the law of the sea

27. Nechaev, *supra* note 25, at 50–51.
28. *See* Boston Globe, Aug. 29, 1965, at 22, col. 1; P. BARABOLIA ET AL., VOENNO-MORSKOI MEZHDUNARODNO-PRAVOVOI SPRAVOCHNIK 289 (1966), *translated as* NAVAL INTERNATIONAL LAW MANUAL (1968).
29. Barabolia in OKEAN, *supra* note 6, at 17–18.

that it formerly had, Soviet jurists nonetheless continue to regard certain bodies of water adjacent to Soviet coasts as having a special juridical status. The latest works return to Dranov's thesis that the legal nature of the straits leading to a particular sea is dispositive.[30] Barabolia identifies five types of strait: (1) straits leading into internal seas or bays that are internal waters of the coastal State; (2) historic straits; (3) straits of archipelagoes; (4) straits leading to closed seas; and (5) international straits.

An example of the first type is the Kerch Strait, linking the Sea of Azov and the Black Sea. The second type has been discussed above. With regard to the third class, Barabolia mentions the Indonesian Declaration of September 13, 1957,[31] (which was expressly endorsed by the Soviet Government), the Ecuadorian decrees of 1938 and 1951 relating to the Galapagos, and "claims" to archipelago waters advanced by the Philippines and Fiji.[32] The fifth category embraces straits linking open seas and oceans or two parts of a single open sea.[33]

30. *See* B. DRANOV, CHERNOMORSKIE PROLIVY: MEZHDUNARODNO-PRAVOVOI REZHIM (1948).
31. Pravda, Feb. 13, 1958, at 2, col. 1. It has been reported recently, though not from Soviet sources, that the Soviet Union is acceding to Indonesian pressure and no longer regards the Strait of Malacca as an international waterway. It is unclear whether Soviet naval vessels will comply with the requirement of prior notification before transit. 15 OCEAN SCIENCE NEWS, No. 1, at 1 (1973). In a February 1973 article, however, the Legal Adviser to the USSR Ministry of Foreign Affairs expressly declared that the Strait of Malacca is an international strait (and the Strait of Tiran is not). Appearing on the eve of the Seabed Committee deliberations in New York, this article can probably be regarded as an authoritative representation of Soviet attitudes. *See* O. Khlestov, *Mezhdunarodno-pravovye problemy Mirovogo okeana,* MEZHDUNARODNAIA ZHIZN', No. 2, at 56 (1973), *translated in* INT'L AFFAIRS [Moscow], No. 3, at 34–44 (1973). On March 27, 1973, the *People's Daily* [Peking] criticized Khlestov's article, particularly its assessment of the attitudes of less developed countries and its posture toward arms-control proposals at sea. For the Soviet reply, see *Mezhdunarodnoe pravo v khunveibinovskii interpretatsii,* MEZHDUNARODNAIA ZHIZN', No. 5, at 128–29 (1973). In February 1973 Admiral Gorshkov also dwelt upon the law of the sea in the conclusion to a long series of articles on the future role of the Soviet navy. He did not mention any particular straits by name, but endorsed the right of free passage through straits connecting high seas and used for international navigation. *See* Gorshkov, *Voenno-morskie floty v voinakh i v mirnoe vremia,* MORSKOI SBORNIK, No. 2, at 17 (1973). For a general discussion of the Strait of Malacca, see Leifer & Nelson, *Conflict of Interest in the Straits of Malacca,* 49 INT'L AFFAIRS [London] 190–203 (1973).
32. For the text of the Philippine Note of Dec. 12, 1955, see 3 SBORNIK RE-GIONAL'NYKH SOGLASHENII I ZAKONODATEL'NYKH AKTOV ZARUBEZHNYKH GOSU-DARSTV PO VOPROSAM MOREPLAVANIIA 251 (1969).
33. Barabolia divides the fifth category into two subgroups: straits not overlapped by the territorial waters of the coastal States and those that are so overlapped. While noting that all vessels have a right of innocent passage through the latter subgroup, he believes that "innocent passage" is inadequate because it permits coastal States

Straits leading to closed seas, the fourth category, are said to be distinctive because they are situated apart from the basic world sea lanes and are important economically only for a small number of coastal States. They are open to foreign merchant and fishing vessels just as the high seas. International practice knows only one limitation: warships of nonlittoral powers are to be excluded. In geographic configuration, closed seas are deeply indented into the mainland or almost completely enclosed by land and linked with the oceans or other seas solely by narrow straits. Barabolia discerns two classes of strait leading to closed seas: (1) straits whose regime is regulated by international agreements, for example, the Black Sea and the Baltic straits, which "repeatedly have been recognized by coastal and a significant number of noncoastal powers as completely closed to the access of warships of noncoastal States";[34] (2) straits with regard to which an international agreement has not been concluded, for example, the Korean, Sangar, Soya [La Pérouse], and Kuril Straits leading into the Sea of Japan and the Sea of Okhotsk.[35] This formulation makes it clear that security considerations are paramount in advancing the closed sea concept and that every effort is being made to shape the geographic and legal criteria so that only seas washing Soviet coasts can be thus classified.

The Continental Shelf

Under any of the proposals enjoying currency in international deliberations, the Soviet Union will be a major continental shelf power.

---

to use or interpret that concept in a unilateral and perhaps arbitrary manner: "The expansion of ties among States dictates the necessity for an international legal guarantee of freedom of navigation through the most important international straits." Barabolia, *supra* note 29, at 25. On July 25, 1972, the Soviet Union submitted to Sub-Committee II of the U.N. Seabed Committee Draft Articles to define freedom of transit through straits used for international navigation. The rules that coastal States and transiting vessels should observe are laid down in considerable detail, but a definition or list of straits used for international navigation is not attempted. That submarines are not specifically singled out implies that they would be permitted to transit such straits while submerged. *See* U.N. Doc. A/AC. 138/SC. II/L. 7 (July 25, 1972); excerpts are reproduced in NEW DIRECTIONS IN THE LAW OF THE SEA 554–56 (S. Lay, R. Churchill, & M. Nordquist eds. 1973) [hereinafter cited as NEW DIRECTIONS].

34. Barabolia, *supra* note 29, at 20.
35. *Id.* at 19–21. *See also* A. VOLKOV, *supra* note 26, at 124, 129. The Kuril Strait, between Kamchatka and the Kuril Islands, is 6 nautical miles at its narrowest point. Soya Strait is 23 miles wide between Hokkaido and Sakhalin (20 miles between Hokkaido and Ostrov kamen' opasnosti; 9 miles on to Sakhalin).

Rough areal allocations projected by the Geographer of the United States Department of State suggest that were a 40–nautical mile limit to be adopted, the Soviet Union would be the third-ranking country with nearly ten percent of the world's shelf area; under a 200–nautical mile limit its shelf area would increase appreciably, but its rank and percentage would decrease to sixth, with 5.3 percent of the world shelf. Under either the 200-meter limit or the edge of the margin limit, the USSR would rank fifth and possess 5.8 or 5.5 percent of the total world shelf.[36]

The breadth of the geological shelf adjacent to Soviet coasts varies from sea to sea. Best endowed are the seas situated in the regions climatically most adverse, the Barents, White, Kara, Laptev, East Siberian, Chukchi and Bering Seas. The shelf of the Baltic Sea has been declared to be continuous and, accordingly, subject to delimitation by agreement among the Baltic States.[37] Except for the bed of the Karakinitsk Bay in its northwestern part, the Black Sea has virtually no shelf at all. The Seas of Japan and Okhotsk also possess comparatively narrow shelves that drop off sharply. The beds of the Caspian and Aral Seas and the Sea of Azov are all unusual cases and not of any great consequence for the international community.

Exploitation of the continental shelf is still at a very early stage in the Soviet Union. With the exception of offshore oil production in the Caspian Sea, which began in 1949 and has severely polluted the sea at enormous cost to fish stocks, most shelf activity is of an exploratory nature. A few pilot projects have been initiated recently. Offshore wells are being drilled near Sakhalin, and there are reports of oil potential in the Baltic and certain Arctic seas, particularly the Kara Sea.[38] Minable deposits of gold may exist in the bed of the Chukchi Sea; placer deposits of tin ore are being mined from a lighter in the Vankina Gulf. There are encouraging reports of complex placer deposits containing cassiterite, magnetite, zircon, and rutile in several

36. *See* Office of the Geographer, Dep't of State, Theoretical Areal Allocations of Seabed to Coastal States Based on Certain U.N. Seabed Committee Proposals, International Boundary Study, Series A, Limits in the Seas, No. 46 (Aug. 12, 1972). The USSR would gain some 18,600 sq. n.m. of shelf under a combined 200-mile–200–meter proposal, but its rank order and percentage of shelf would not be affected.

37. *See* the Declaration of the Continental Shelf of the Baltic Sea, *signed* at Moscow, Oct. 23, 1968, by the USSR, Poland, and the German Democratic Republic, Izvestia, Oct. 24, 1968, at 3, col. 1, 6 SOV. STAT. & DEC. No. 3, at 261 (1970).

38. For a report of natural gas in the Kara seabed and predictions of vast petroleum deposits underlying the Arctic seas, see SOVIET NEWS, Feb. 3, 1970, at 58. But the resource potential of the Arctic shelf is virtually ignored in a recent monograph on the Soviet Arctic. *See* S. SLAVIN, THE SOVIET NORTH 186 (1972).

areas of the Arctic seabed. In the Baltic, a pilot mining project operated from 1968 to 1970 demonstrated that rare minerals could be extracted more cheaply than by equivalent activity on land. Exploratory expeditions have identified placer deposits of gold and tin in the Sea of Japan, as well as phosphorites whose extraction may be less expensive for the Far East than fertilizers imported from Soviet Central Asia. Preliminary estimates suggest that rutile, zircon, titanium-magnetite, cassiterite, magnetite, tungsten, gold, and diamonds can be profitably mined offshore of the USSR.[39]

It is highly unlikely that the Soviet shelf, technology permitting, will be the object of large-scale exploitation until at least the end of the decade. The current Five-Year Plan places no particular stress on shelf exploration, and the Directives of the Twenty-fourth Congress of the Communist Party of the Soviet Union call only for an expansion of "studies of coastal placer deposits of gold, tin, and other ore deposits."[40] By the end of 1975 Soviet marine economists hope to have a scientifically substantiated forecast of the volume of useful deposits on the continental shelf of the USSR plus the methodology of searching for and working them. They then expect to begin the task of creating a material and technical base so that a large-scale mineral raw-materials industry could begin on the bed of the seas and oceans during the following five-year period.[41]

Although Soviet jurists have not taken a firm position on what the seaward limit of the continental shelf should be, the tenor of their writings suggests they would be inclined to choose a narrow shelf. They reject the principle of exploitability out of hand as the sole criterion for delimiting the shelf:

The history of the formation of the international legal norm on the continental shelf permits one to draw the conclusion that sovereign rights of the coastal State for the purpose of exploring and exploiting the natural resources of the continental shelf cannot be extended for an indefinite distance seaward from the outer boundary of its territorial waters. The criteria of a 200-meter depth and "exploitability" mutually supplement one another. The spatial ex-

---

39. Mikhailov, *Razvitie morskoi ekonomiki SSSR,* Voprosy Ekonomiki, No. 7, at 101–08 (1972).

40. Gosudarstyennyi piatiletnii plan razvitiia narodnogo khoziaistva sssr na 1971–1972 gody (1972); Materialy xxiv s'ezda kpss 253 (1971). An article released by Novosti Press Agency concluded with the observation that "the country's leaders feel that the future must definitely incorporate the resources from the sea and scientists and technicians are committed to bringing marine resources to large-scale commercial production within the next ten to twenty years." *See How the U.S.S.R. Plans to Develop Its Ocean Resources,* Ocean Industry, No. 12, at 26 (1970).

41. *See* Mikhailov, *supra* note 39.

tension of the "exploitability" criterion is limited by the requirement that a submarine area be adjacent to the coast of the particular State or island.[42]

Similarly, it is doubtful, in the Soviet view, that the 1958 Convention on the Continental Shelf embraces the continental slope.[43] Soviet jurists point out that proposals variously made by Panama, England, and the Netherlands expressly to link the continental slope with the shelf were not accepted. According to Smirnov, "[I]t is obvious from this that the conference participants were not inclined to include the slope in the juridical concept of the continental shelf."[44] He speculates that an international agreement on a more precise seaward limit of the shelf is likely to shun exploitability considerations in favor of a suitable combination of depth and adjacency, having due regard for the interests of States lacking a geological shelf.

The Deep Seabed

During 1970–72 Soviet attitudes toward a future seabed regime crytallized somewhat in comparison with the earlier period, although they remain highly fluid, as do those of the other major powers. Soviet scientists continue to express optimism about the resource potential of the deep seabed, while remaining skeptical about large-scale commercial exploitation at great depths in the near future.[45] Various legal regimes for the seabed are being discussed and evaluated in Soviet legal writing with greater assurance and sophistication than before. The bare outlines of a juridical framework that the USSR would like to see were laid down on July 22, 1971, in Provisional Draft Articles of a Treaty on the Use of the Sea Bed for Peaceful Purposes submitted to the United Nations Seabed Committee.[46]

Legal Conceptions of the Seabed

Lazarev has identified five distinct legal conceptions on which exploitation of the seabed beyond the continental shelf might be based:

42. Smirnov in OKEAN, *supra* note 6, at 97. However, Smirnov acknowledges that some departure from a purely geological concept of the shelf obviously is justified in order to grant equivalent rights to States not possessing a geological shelf, *e.g.*, in the Persian Gulf and off the western coast of Latin America.
43. *Done* Apr. 29, 1958, [1964] 1 U.S.T. 471, T.I.A.S. No. 5578, 499 U.N.T.S. 311.
44. Smirnov, *supra* note 42, at 97. *See also* Tsarev, *Mezhdunarodno-pravovye voprosy Kontinental'nogo Shel'fa*, in AKTUAL'NYE PROBLEMY, *supra* note 17, at 51.
45. *See, e.g.*, Fedinskii, *Problema mineral'nykh resursov dna morei i okeanov i zadachi morskoi razvedochnoi geofiziki*, SOVETSKAIA GEOLOGIIA, No. 5, at 4 (1969).
46. 11 INT'L LEGAL MATERIALS 778 (1972).

(1) the "legal vacuum" approach; (2) *res nullius;* (3) *res communis;* (4) common heritage of mankind; and (5) joint or common use.[47]

The Legal Vacuum Approach

The ocean floor beyond the continental shelf as an expanse to which prevailing principles of international law do not extend has been rejected from the outset as simply contrary to fact. The overwhelming majority of States and scholars are said to acknowledge that the United Nations Charter and modern international law apply there, just as on earth, in outer space and in the atmosphere. The legal vacuum view is held to be dangerous because it offers "certain States and their ruling classes" the opportunity to reject "democratic prevailing principles of contemporary international law."[48] Admittedly, there are few such norms since the deep seabed has not been exploited to any great extent and those that do exist were "worked out as part of the international legal regime of the high seas."[49]

There is no consensus yet as to precisely what principles of international law do apply to the seabed. The United Nations Charter is regarded by all Soviet jurists as being applicable, and several believe that certain special international legal norms also extend to the area. Ostrovskii[50] mentions those articles of the 1958 Convention on the High Seas relating to radioactive substances,[51] the 1963 Nuclear Test Ban Treaty,[52] and the 1959 Treaty on Antarctica.[53] To this list Klimenko adds Article 13 of the 1958 Geneva Convention on Fishing[54] and "imperative, generally recognized principles of international law," such as sovereign equality, prohibition of the threat or use of force, and peaceful settlement of disputes.[55] But Lazarev points out that the number of special international legal norms regulating utilization of the seabed are few because, in comparison with the superjacent water column, there has been little seabed activity. What norms do specially apply, he adds, were developed as part of the legal regime of the high

47. Lazarev in OKEAN, *supra* note 6, at 116.
48. *Id.* at 116–17.
49. Kalinkin in OKEAN, *supra* note 6, at 132.
50. Ostrovskii in MORSKOE DNO: KOMU ONO PRINADLEZHIT? 104 (G. Kalinkin & Ia. Ostrovskii eds. 1970).
51. *Done* Apr. 29, 1958, [1962] 2 U.S.T. 2312, T.I.A.S. 5200, 450 U.N.T.S. 82.
52. *Done* Aug. 5, 1963, [1963] 2 U.S.T. 1313, T.I.A.S. 5433, 480 U.N.T.S. 43.
53. Multilateral Antarctic Treaty, *done* Dec. 1, 1959, [1961] 1 U.S.T. 794, T.I.A.S. No. 4780, 402 U.N.T.S. 71.
54. Convention on Fishing and Conservation of Living Resources of the High Seas, *done* Apr. 29, 1958, [1966] 1 U.S.T. 138, T.I.A.S. 5969, 559 U.N.T.S. 285.
55. Klimenko in AKTUAL'NYE PROBLEMY, *supra* note 17, at 74.

seas. Except for the resources of the shelf zone proper, the seabed has never been considered separately from the superjacent waters.[56] Since, according to Lazarev, the deep seabed is an "integral constituent element" of the concept of the high seas, the seabed cannot be an object of national appropriation.[57] It is available to the common use of all countries and peoples, and accessible as a matter of right to all States that wish to take advantage of its resources.

Res Nullius *and* Res Communis

Res nullius is rejected by Soviet jurists on several grounds. Some see it as an historical anachronism, appropriate in the early days of shelf exploitation when effective occupation of small areas in no way interfered with traditional use of the sea for commerce and trade. Applied to the deep seabed, *res nullius* would be incompatible with the freedom of the seas.[58] Others believe *res nullius* is "scientifically unfounded" because it confuses the legal relationships of sovereign States in respect to maritime law with private civil law relationships of the law of things. More damning, perhaps, is the further observation that "politically the said concept is dangerous because it tries to legalize seizure of the seabed by individual States or a group of States. . . ."[59] *Res communis* is criticized on analogous grounds.

The Common Heritage of Mankind

The concept of the seabed as the common heritage of mankind has been the object of a multitude of criticisms by Soviet jurists. Initially it was linked with *res communis* or with the creation of an international mechanism. Although Soviet opposition to the latter has mellowed in principle, and the notion of a common heritage of mankind is said to contain a rational core insofar as it stresses the equal rights of all to an international seabed territory and the inadmissibility of claims by some States to the detriment of others, Lazarev nonetheless views it as politically dangerous and unscientific. It seeks to replace the true subjects of international relations—States—by focusing

56. Lazarev, *supra* note 47, at 133–34. Lazarev cities with approval Henkin's observations on the same issue from L. HENKIN, LAW FOR THE SEA'S MINERAL RESOURCES 18 (1968).
57. Lazarev, *supra* note 47, at 135.
58. A. ZHUDRO & A. KOLODKIN, LEGAL ASPECTS OF USING THE SEABED 10–17 (1969).
59. Lazarev in OKEAN, *supra* note 47, at 117.

on an "amorphous, apolitical mass of 'mankind.' "[60] The concept is said to be politically dangerous in that, under the guise of introducing true equality into international maritime relations (which, Lazarev observes, for a number of historically determined reasons has never been the case on land), it does so at the expense of the technologically advanced States. All of the technologically advanced powers, whether capitalist or socialist, Lazarev complains, would be thrown together in a single group; the common heritage presumes a division of the world into poor and rich nations, irrespective of their class structure.[61]

Joint or Common Use

Soviet jurists find the principle of joint or common use of the seabed to be preferable. Derived from the principle of freedom of the seas, it holds: (1) all States and peoples are entitled to exploit the seabed of the "world ocean" beyond the seaward limit of the continental shelf; and (2) no single State can claim any exclusive rights and privileges whatever to the detriment of other States and peoples. It is recognized that conflicts over competitive use will arise, as, for example, when two or more States wish to exploit a single parcel adjacent to a deposit being worked by others or when one or several States desire to use a segment of water column for traditional activity, such as navigation or fishing, and others for nontraditional activity, such as mining or scientific research. Consequently, the principle of common use is said to presuppose the establishment of a particular international regime for the deep seabed and possibly of an international mechanism with a view toward averting violations of the principle of equal use and facilitating the observance of international law.[62]

Soviet acknowledgement that an international organization may be required is a significant modification of previous attitudes on the question. Juridical literature published in the USSR to date, however, has given little attention to the issue, dwelling instead on proposals advanced by Western Governments or marinists. But Uustal's commentary,[63] coupled with the provisional draft articles submitted to the

---

60. *Id.* at 118-19. Lazarev sees proposals for an ocean development tax on the leading industrial powers as an extension of the common heritage theory; he is especially apprehensive that such a tax might be applied to all high seas activities.
61. *Id.* at 118.
62. *Id.* at 119-20.
63. Uustal', *Komu prinadlezhat okeanskie bogatstva?*, 6 SOVETSKOE PRAVO [Tallin] 280-84 (1972).

Seabed Committee by the USSR, casts some light on what seabed regime the Soviet Union would prefer to see develop.

Legal Status and Limits of the Seabed

Soviet attitudes toward the principal seabed regimes have been outlined above. The Draft Provisional Articles can be said to embody the theory of common use, although the term is not explicitly invoked.[64] The seabed and subsoil thereof, under Article 1 of the Provisional Draft, would be "open to use exclusively for peaceful purposes to all States, coastal or landlocked." No State is to claim or recognize the claim of another to any part of the seabed or subsoil, nor is the seabed to be subject to appropriation by any means, States, or persons, natural or juridical, according to Article 5. Articles 25 and 26 provide that any rights granted to, or exercised by, a Seabed Resources Agency created pursuant to the provisional draft shall not be construed as jurisdiction over the seabed, the superjacent waters, or airspace. Pending international agreement on the seaward boundary of the continental shelf, Article 3 of the provisional draft leaves open the limits of the seabed.

Use of the Seabed

The Provisional Draft lays down a number of principles in respect to the use of the seabed. Article 4 provides that exploration and exploitation of seabed resources are not to conflict with the principles of freedom of navigation, fishing, research, and other activities on the high seas. Under Article 7 States are to act in regard to the seabed in conformity with principles and rules of international law, including the United Nations Charter, the 1970 Declaration on Principles of International Law Concerning Friendly Relations and Cooperation Among States[65] and the 1960 Declaration on the Granting of Independence to Colonial Countries and Peoples.[66]

Article 8 provides that commercial exploration and exploitation of seabed resources is to be carried out for "the benefit of mankind as a whole," taking into particular consideration the needs of the de-

64. Provisional Draft Articles of a Treaty on the Use of the Seabed for Peaceful Purposes. 11 INT'L LEGAL MATERIALS 778 (1972). All references to the provisional Draft hereinafter are to this text.
65. G.A. Res. 2625, 25 U.N. GAOR Supp. 28, at 121, U.N. Doc. A/8082 (1970).
66. G.A. Res. 1514, 15 U.N. GAOR Supp. 16, at 66, U.N. Doc. A/4684 (1960).

veloping countries. Article 9 envisages some sort of commercial licensing scheme, but the details, as in the case of distribution of benefits under Article 14, are not given.

Soviet jurists dismiss outright proposals to impose an international levy on a State's activity within its own territorial and internal waters as a form of financial and political interference in internal affairs. The Soviet view is that it is the prerogative of every State to establish its own system of finance and taxation, the refusal to permit the levy of foreign taxes on its territory being deemed an inalienable attribute of State sovereignty.

As for a levy on revenues from, or activity on, the deep seabed, the views expressed so far are equivocal. A tax on every ocean activity for the purpose of stimulating the future development of ocean exploitation would, in the Soviet view, be regressive and actually discourage investment in seabed operations. Moreover, the traditional maritime powers, which rely heavily on the sea for fish or minerals, would be placed at a disadvantage, although former metropolitan States such as Great Britain might justifiably be taxed to settle "historical accounts." But even this "debt," Soviet jurists stress, is a moral and political one, not a legal obligation.[67]

The abuse of taxation as a means to political ends is a recurrent theme in Soviet writings.[68] Small States that wish to redress disparities in national power through an ocean tax are dismissed as being utopian. The USSR emphasizes that it will not be coerced into extending aid to smaller countries, nor will it acknowledge a legal obligation to do so. Such assistance, Soviet jurists maintain, has been granted in the past for humane reasons. Small States hungry for revenues ought first to tax foreign firms operating on their continental shelves before looking to the international community.

An international charge on State activity within an international zone and connected with exploitation of seabed resources is not wholly excluded, however. Lazarev comments: "Sovereign States may come to some general agreement on the basis of equality and mutual advantage, observing the principle of universality."[69]

Some clue as to what might be acceptable in principle to the Soviet Union by way of such an arrangement is offered in Lazarev's cautious endorsement of a scheme to employ financial sanctions against States for polluting the ocean. Sanctions of this type apparently would be seriously considered, provided that States would voluntarily consent

67. Lazarev in OKEAN, *supra* note 47, at 124.
68. *See, e.g., id.* at 121–25.
69. Lazarev, *supra* note 47, at 122.

to the possibility of being fined, that each State would bear equal liability for pollution irrespective of its level of economic development, and that the sanctions would in no way appertain to the international taxation of State activity within territorial or internal waters.

Seabed Installations

As a maritime power, the USSR has been concerned that an appropriate balance be struck among seabed exploitation and other uses of the sea. The general assertion of this proposition in Article 4 of the Provisional Draft already has been alluded to in this connection.[70] Article 10 of the Draft stipulates that stationary and mobile installations may be erected or emplaced with a view to commercial exploration and exploitation of the seabed, on condition that they are not placed on straits, at points where they may obstruct international shipping lanes, or at points of intensive fishing activity. Safety zones analogous to those around continental shelf installations, each 500 meters in radius, would be established, but they could not be placed so as to form a belt barring the access of shipping to particular zones or cutting across international sea lanes. Appropriate notice of underwater installations would be given to the international community.

In rather curious wording, paragraph 4 of Draft Article 10 seems to contemplate that *States* will erect and emplace installations within sections allotted to them and dismantle them when their allotment period expires, unless they are acquired by another State. Installations are not to have the status of islands or their own territorial sea, nor are they to affect the delimitation of the territorial sea or the seabed.

Under Article 12 the dimensions and configuration of seabed sectors being exploited are not to be situated so as to form a belt across zones through which the vessels of States having no coastline on the Atlantic, Pacific, or Indian Oceans make their way to the waters of these oceans or to the international sea lanes crossing them. In the Soviet Union, the avoidance of such obstructions is essential for access to the Atlantic Ocean and to some areas of the Pacific Ocean.[71]

Seabed Resources Agency

At the outset of seabed deliberations, Soviet diplomats were implacably opposed as a matter of principle to the United Nations Sec-

70. *See* text accompanying notes 65 and 66 *supra*.
71. *See* Klimenko, *supra* note 55, at 82.

retariat even undertaking a study of possible forms of international machinery to administer a seabed regime. Such machinery, it was insisted, would "serve exclusively the interests of capitalist monopolies."[72] Presumably the specter of unfettered competition among nearly a hundred coastal States was instrumental in persuading the Soviet leadership to reconsider its position. Whatever the cause, Soviet jurists no longer reject proposals for an international seabed administration out of hand. Nonetheless, almost all of these proposals continue to be unacceptable for one of two reasons. Either they are supranational in nature, infringing the sovereign rights of States or they seek to divide the international community into two types of States: those that work the sea and those that reap the fruits of the sea.

The Provisional Draft Articles outline the framework of an international mechanism that the Soviet Union evidently would be prepared to accept. An International Sea-Bed Resources Agency would be established, consisting of all States parties to the seabed treaty. The Agency would meet every two years (unless an extraordinary session were convoked) as the "Conference of the Agency" for the purpose of establishing the Executive Board, considering and approving the Agency's administrative budget, considering general questions relating to the exploitation of seabed resources, considering reports of the Executive Board, appointing the executive secretary of the Agency, drafting general principles and recommendations to States concerning the preventing of pollution and contamination of the marine environment resulting from seabed exploration and exploitation, and considering other issues that might arise in the application of the treaty unless they fall within the jurisdiction of the Executive Board. In addition, the Conference might adopt resolutions on the recommendation of the Executive Board, or the United Nations Security Council, to deprive States of membership in the Agency. Each Conference member would have one vote. Procedural decisions would be taken by a simple majority, substantive decisions by a two-thirds majority of members present and voting according to Article 18.

Between regular sessions of the Conference, the Executive Board would meet at least once annually to carry out the functions delegated to it by the treaty. The proposed Board would be dominated by the developing countries. It would consist of thirty States, six (including one landlocked power) to be elected to a four-year term from each of the following groups of countries: (a) socialist; (b) Asian; (c) African; (d) Latin American; (e) Western European and others not falling within

72. U.N. Doc. A/PV. 1752 (Dec. 21, 1968) (Mendelevich).

the other categories. As provided by Article 22, the Executive Board
would: (1) supervise the implementation of the treaty by member
States and activities connected with the commercial exploration and
exploitation of seabed resources; (2) coordinate the activities of States
parties in commercial exploration of seabed resources and make a
general evaluation, based on data supplied by States, of the reserves of
proven resources, their distribution on the seabed, and the depth at
which they occur; (3) supervise compliance with certain treaty provisions; (4) consider specific problems arising for landlocked countries
in connection with seabed exploration and exploitation; (5) promote
exchanges of related scientific and technical information; (6) adopt
recommendations on ways to prevent pollution of the marine environment and damage to living resources of the sea resulting from commercial seabed exploration and exploitation; (7) assist in settling
disputes between States concerning implementation of the treaty; (8)
establish for the settlement of disputes, at the request of parties
thereto, organs of conciliation, arbitration, and so forth; (9) consider
other issues arising out of treaty provisions; and (10) carry out as yet
unspecified functions in regard to licenses and the distribution of
benefits. Decisions of the Executive Board would be taken in the same
way as those of the Conference of the Agency. States not represented
on the Board might participate in the Board's deliberations, without a
right to vote, if the issue being considered directly affected their
interests.

The powers that would devolve on the contemplated Conference
and Executive Board are not far-reaching; nor is a proposed Secretariat for the administrative and technical servicing of the Agency and
its organs to assume a role of major consequence. The Agency may
perhaps be described as more akin to a conference of States than to an
organization of States. As before, the Soviet Union continues to be
reluctant to support the creation of an international organization
merely because that seems the expedient action to take; in the Soviet
view, it must be a *necessary* measure. Klimenko expresses this view
when he writes: "But the real necessity for the creation [of an international agency] arises only with the developing exploitation of the
seabed in practice. Moreover, the creation of an international agency
depends on the preliminary resolution of other questions, particularly
the issue of the basic principles on which international legal regulation
of seabed exploitation should be based."[73]

Soviet critiques of international organizational arrangements
proposed by other powers, by public and private groups, and even by

73. Klimenko, *supra* note 57, at 83.

individual citizens cast some light on what is objectionable in principle and what is merely undesirable. The USSR is unwilling to delegate to an international agency either the power to decide whether a given State may explore or exploit the seabed or subsoil thereof or the authority to carry out scientific research. Nor is it willing to place the technologically advanced States at the mercy of less-developed countries in matters of seabed or ocean revenues. Apprehension is expressed that eventually all ocean activities, including merchant shipping and even naval operations, could be regarded as a source of revenue to a majority of underdeveloped countries.[74] Proposals from American sources favoring internationalization of the seabed are seen as a mask designed to ensure a carry-over of "methods of the capitalist system of management"[75] to the seabed, although these capitalist techniques are not specified.[76]

Kalinkin attributes the United States proposal for an international trusteeship zone to the "American oil monopolies."[77] By using "various forms of pressure" and making appropriate payments to coastal States "without harm to themselves,"[78] American petroleum companies are said to have already secured exclusive rights to explore and exploit seabed resources in the coastal waters of Asia, Africa, and Latin America. He acidly observes that those less-developed countries most generous in granting concessions to foreign monopolies are the ones that advocate most vocally a system of international control over seabed development.[79]

Lazarev points out that, absent a treaty, there are no juridical obstacles to a State's establishing provisional jurisdiction over areas of seabed beyond the continental shelf it it is technologically capable of exploiting them. Without a special treaty these expanses of seabed are in the "common use of all States and peoples."[80] Jurisdiction over a particular area may be passed from one State to another, Lazarev maintains, and the principle of freedom of the seas will reign supreme.

74. Lazarev, *supra* note 49, at 120, 125. Kalinkin sees the original Maltese proposal of Aug. 17, 1967 as being of this type. Kalinkin, *supra* note 49, at 147.

75. Kalinkin, *supra* note 49, at 138.

76. *Id.* at 138–46. The author's criticism of American proposals is rather general. For the most part, he is content to summarize some of their salient features. Among those examined are proposals of the Commission to Study the Organization of Peace, the World Peace Through Law Center, *Pacem in Maribus,* the American Assembly, the U.S. Commission on Marine Sciences, and Senator Pell of Rhode Island.

77. *Id.* at 57.

78. *Id.*

79. *Id.*

80. Lazarev, *supra* note 57, at 128.

In his view this is not necessarily a bad result. Creating legal rules in anticipation of, rather than in response to, technological progress could, he argues, be harmful and unjust for the States concerned and for scientific development. A legal rule once adopted is exceedingly difficult to modify. Years may pass before the requisite unanimous consent of States is achieved. Lazarev urges the international community to defer law-making activity with regard to the seabed until technology points the way. In the meantime, developing countries can participate in seabed exploitation by concluding bilateral or multilateral agreements with advanced States, "taking into account the contribution these [developing] countries may make . . . to exploiting the subsoil and other riches of the World Ocean."[81]

Demilitarization of the Seabed

Soviet seabed diplomacy continues to stress the importance and desirability of demilitarizing all areas of the seabed. The USSR is a party to the 1963 Nuclear Test Ban Treaty,[82] the Nuclear Weapons Nonproliferation Treaty,[83] and the 1959 Treaty on the Antarctic[84]— all of which are regarded as constructive steps toward demilitarization. The Soviet Government played a leading role in initiating and drafting the 1971 Treaty on the Prohibition of the Emplacement of Nuclear Weapons and Other Weapons of Mass Destruction on the Seabed and the Ocean Floor and in the Subsoil Thereof now in force.[85] In May 1972 the United States and the USSR concluded three agreements limiting the deployment of antiballistic missile systems that also appertain to the seabed and superjacent water column.[86]

81. *Id.* at 129.
82. *Done* Aug. 5, 1963, [1963] 2 U.S.T. 1313, T.I.A.S. 5433, 480 U.N.T.S. 43, 23 SBORNIK DEISTVUIUSUCHIKH DOGOVOROV, SOGLASHENII I KOVENTSII, ZAKLIUCHENNYKH SSSR S INOSTRANNYMI GOSUDARSTVAMI 44 (1970) [hereinafter cited as SDD].
83. *Done* July 1, 1968, [1970] 1 U.S.T. 483, T.I.A.S. 6839, [1970] 14 Ved. Verkh. Sov. S.S.S.R. Item 118.
84. *Done* Dec. 1, 1959, [1961] 1 U.S.T. 794, T.I.A.S. 4780, 402 U.N.T.S. 71, 22 SDD 233.
85. [1972] 30 Ved. Verkh. Sov. S.S.S.R. Item 257, 10 INT'L LEGAL MATERIALS 145 (1971).
86. *See* the Treaty on the Limitation of Antiballistic Missile Systems, the Interim Agreement on Certain Measures with Respect to the Limitation of Strategic Offensive Arms, and the Protocol to the Interim Agreement, all signed at Moscow, May 26, 1972. [1972] 45 VED. VERKH. SOV. S.S.S.R. Items 420, 421. English texts are reproduced in NEW DIRECTIONS, *supra* note 33, at 292–306.

Articles 1 and 5 of the Provisional Draft would ensure that the seabed is "open to use exclusively for peaceful purposes by all States" and prohibited "for military purposes." Precisely how this language, were it to be accepted, would constrain naval activities of the principal maritime powers had not been commented on in Soviet legal writing. Kalinkin, however, characterizes complete demilitarization as the "most urgent seabed issue."[87]

Declaration of Principles Regarding the Seabed

On December 17, 1970 the United Nations General Assembly approved a Declaration of the Principles Defining the Bed of the Seas and Oceans and their Subsoil Beyond the Limits of National Jurisdiction by an overwhelming vote.[88] With some misgivings, the USSR was among fourteen States abstaining. The Declaration is treated in Soviet doctrinal literature as being a positive, but far from adequate, basis from which a future seabed regime should emerge.[89] The juristic nature of General Assembly resolutions has been, and continues to be, debated in Soviet legal writing.[90] In commenting on the Declaration, Kalinkin, like other Soviet maritime lawyers, stresses its "recommendatory character," emphasizing that it "cannot be regarded as a legally binding act."[91] He describes certain stipulations in the Declaration—that the seabed beyond the limits of national jurisdiction cannot be appropriated by States or persons, natural or juridical, and that no State can either claim to exercise or actually exercise sovereignty or sovereign rights over any area of the seabed—as "key[s] to the legal status of the seabed."[92]

But he observes that there are "inconsistent, contradictory"[93] provisions whose inclusion in the Declaration are "an unjust attempt to consolidate a unilateral approach"[94] to the legal problems of the

87. Kalinkin in Okean, *supra* note 6, at 132.
88. G.A. Res. 2749, 25 U.N. GAOR Supp. 28, at 24, U.N. Doc. A/8028 (1970), 10 Int'l Legal Materials 220 (1971).
89. For Soviet views on principles relating to the seabed expressed prior to the Declaration, see U.N. Doc. A/AC. 135/SR. 12, at 12–13 (1968); U.N. Doc. A/AC. 138/ SC. 1/8 (Aug. 21, 1969); U.N. Doc. A/AC. 138/SR. 22, at 7–8 (1970).
90. For a summary of various doctrinal positions, see Ianovskii, *Sovetskaia nauka o iuridicheskoi sile rezoliutsii general'noi assamblei OON*, Sovetskiiezhegodnik mezhdunarodnogo prava 1964–65, at 111–21 (1966).
91. Kalinkin, *supra* note 51, at 161.
92. *Id.* at 163.
93. *Id.*
94. *Id.*

seabed. According to Kalinkin, the Declaration "ignores" the crucial importance of fixing a precise limit for the continental shelf, relegating the issue to the Preamble. He notes that while the Declaration provides that the seabed should be used exclusively for peaceful purposes, it fails to make clear that any military activity on the seabed should be proscribed, thereby "not eliminating the possibility that the meaning of the said principle will be distorted by imperialist powers to legalize their activity in using the seabed for military purposes."[95] He contends that some provisions of the Declaration proceed from the premise that coastal States enjoy special rights with regard to the deep seabed, a premise "that finds no confirmation in the contemporary international law of the sea."[96] Furthermore, he believes the Declaration avoids the issue of freedom of scientific research, "a generally recognized international legal principle."[97]

Kalinkin deplores the references in the Declaration to the seabed as the common heritage of mankind as "demagogic," "politically unreal" attempts to equate two systems of ownership that are inherently incompatible.[98] As a socialist State, he argues, the Soviet Union cannot be a "co-owner" of "collective property"[99] side by side with capitalist States, the more so if some co-owners may lawfully use the seabed for purposes antithetical to Soviet interests.

Kalinkin also finds objectionable and "unjustified"[100] the provision in the Declaration regarding the "interests and needs of the developing countries"[101] in apportioning the proceeds from seabed operations. This issue cannot be resolved, he maintains, without taking into account "the socio-political structure of the contemporary world."[102] He says it would be only just to deduct a percentage of seabed revenues from former colonial powers or capitalist monopolies, since they are responsible for the economic backwardness of the developing countries, but it would be unfair "to carry over to the sphere of legal regulation of States' activity in exploiting the seabed an indiscriminate division of the world into a 'rich north' and a 'poor south,' imposing a joint liability for the economic backwardness of the developing countries on both the imperialist colonial powers and the socialist coun-

95. *Id.* at 164.
96. *Id.* at 164–65.
97. *Id.* at 165.
98. *Id.*
99. *Id.* at 167.
100. *Id.*
101. *Id.*
102. *Id.*

tries, which in no way participated in colonial or neocolonial exploitation."[103]

Scientific Research on the Seabed

Soviet jurists affirm that one of the freedoms of the seas is the freedom to carry on scientific research. It applies "to the high seas as a whole, including the seabed."[104] The Provisional Draft Articles stipulate that neither the treaty nor any rights granted or exercised pursuant thereto shall affect the freedom of research on the seabed or the subsoil thereof. To further effective seabed exploitation, Article 27 of the USSR proposal would encourage international cooperation among scientists of different countries that engage in seabed research, publish programs and disseminate the results of research through international and other channels, and cooperate in expanding the research facilities of developing countries, including the development of measures to increase the participation of nationals from such nations in research.

A working paper on international cooperation in marine scientific research, submitted August 3, 1972 to Sub-Committee III of the United Nations ‚Seabed Committee by Bulgaria, the Ukrainian S.S.R., and the Soviet Union, carries considerably further the principles laid down in the Provisional Draft Articles.[105] Such research, the paper recommends, should be conducted solely for peaceful purposes and for the benefit of all countries in accordance "with universally recognized principles and norms of international law."[106] It should advance knowledge of all natural processes and phenomena in the ocean, including the seabed and ocean floor. Coastal States would be obliged to simplify access to ports for marine research vessels and cooperate in providing favorable conditions for marine research. The working paper foresees a role for the International Oceanographic Commission. Member States would also undertake to cooperate in preventing hindrances to the normal functioning and safe preservation of

103. *Id.* at 169.
104. Klimenko, in AKTUAL'NYE PROBLEMY, *supra* note 55, at 89.
105. U.S. Doc. A/AC. 138/SC. III/L. 23 (Aug. 3, 1972). Scientific research does not embrace industrial prospecting, in the Soviet view. The Ukrainian delegate, Kolesnikov, has commented on the "difference of a legal nature between the two types of operations." U.N. Doc. A/AC. 138/SC. III/SR. 10 (June 25, 1971) (Kolesnikov).
106. *Id.*

scientific equipment and installations on the high seas. By the same token, scientific research would have to be conducted without causing ecological damage and without interference to navigation or fishing. These principles, the working paper makes clear, would extend to both the continental shelf and the deep seabed.

Curiously, the entire subject of scientific research at sea has received little attention from Soviet jurists. This continues to be the case despite an increasingly deep Soviet commitment to marine research.

The High Seas

Prevention of Pollution

Control of marine pollution remains a prominent concern in Soviet legal writing. In September 1971 the Soviet Government followed the example of Canada, taking measures to protect the Arctic from the threat of pollution by passing vessels. The USSR Council of Ministers established a Northern Sea Route Administration attached to the Ministry of the Maritime Fleet "for the purpose of ensuring the safety of Arctic navigation, as well as of taking measures to prevent and eliminate the consequences of pollution of the marine environment and the northern coasts . . . ." of the USSR[107] The drafting and implementation of antipollution requirements (including minimum technical standards for vessels intending to transit the Northern Sea Route), the suspension of navigation in areas where pollution may be a problem, and the imposition of penalties have been delegated to the new Administration.

Measures to reduce pollution in the Baltic Sea have been under active discussion among the Baltic powers since September 1969, when they met at Brussels to seek a multilateral approach to the problem. Soviet jurists strongly support a regional solution.[108] The recent normalization of relations with the German Democratic Republic having removed a major political obstacle on the part of Western Governments, seven Baltic States concluded the Convention

---

107. Statute of the Administration of the Northern Sea Route, 17 Sobranie postanovlenii soveta ministrov S.S.S.R., Item 124 (1971) (Council of Ministers USSR), 11 INT'L LEGAL MATERIALS 645 (1972). For background on the Statute, see Butler, *Pollution Control and the Soviet Arctic,* 21 INT'L & COMP. L.Q. 557–60 (1972); *id., Soviet Maritime Jurisdiction in the Arctic,* 16 THE POLAR RECORD 418–21 (1972).

108. *See, e.g.,* Nekrasova in OKEAN, *supra* note 6, at 222.

on Fishing and Conservation of the Living Resources in the Baltic Sea and the Belts in 1973.[109]

Military Exercises

On May 25, 1972, the Soviet Union and the United States signed an Agreement on the Prevention of Incidents on and over the High Seas[110] with a view to clarifying some of the Rules of the Road laid down in the 1960 International Regulations for Preventing Collisions at Sea.[111] Naval vessels of both the Soviet and American fleets had previously been involved in incidents or collisions resulting in diplomatic protests. Admiral Gorshkov has stressed that the Agreement "corresponds fully to the spirit and letter of the 1958 Geneva Convention on the High Seas and rests on the main principles of this Convention, which affirms the right of freedom of navigation for warships of all States on the waters of the high seas."[112] He observes that although the "Agreement in no way infringes the interests of any third States," being bilateral, "if any State expresses a desire to accede thereto, this would only be welcomed."[113] According to Gorshkov, appropriate orders and regulations in pursuance of the Agreement have been issued to Soviet naval personnel.[114]

Commercial Fishing on the High Seas

Soviet legal doctrine has had little new to say during the past two years about fishing on the high seas.[115] Unilateral claims by States to fishing zones in excess of twelve nautical miles continue to be deplored as contrary to international law. The futility of enlarged fishing zones is demonstrated by observing that catches are not believed to have increased for American fishermen after the United States created a

109. *Done,* Sept. 13, 1973, INT'L LEGAL MATERIALS 1291 (1973). *See also, e.g.,* 3 WORLD ECOLOGY 2000, No. 25, at 4 (1972).
110. *Done* May 25, 1972, 11 INT'L LEGAL MATERIALS 778 (1972).
111. *Done* June 17, 1960, [1965] 1 U.S.T. 794, T.I.A.S. 5813.
112. Interview with Admiral S. Gorshkov, "Za bezopasnost' plavaniia v otkrytom more," Izvestia, July 8, 1972, at 3, col. 3. For an expert discussion of the definition and prerogatives of a warship under international law, see Ivanashchenko, *Svoboda otkrytogo moria i problemy voennogo moreplavaniia,* in AKTUAL'NYE PROBLEMY, *supra* note 17, at 91–141.
113. *Id.*
114. *Id.*
115. *See, e.g.,* Volkov in OKEAN, *supra* note 6, at 207.

twelve-mile fishing zone. Volkov decries discriminatory measures directed against Soviet fishing vessels in the Pacific by Canada, Mexico, and Australia: "The goal of these measures is the aspiration to maximize the difficulty of supplying the Soviet fishing fleet with fresh water, fuel, and provisions, to reduce the economic efficiency of its use, and to create hindrances for the loading and unloading work that our vessels working in the eastern Pacific carry out almost exclusively on the high seas."[116] States such as Argentina, Volkov adds, which create exceedingly broad territorial seas, are contributing to a "chronic underutilization" of fishery resources and consequently to a "senseless destruction of enormous biological resources."[117]

Following earlier practices, the Soviet Union endeavors to come to terms with coastal States that seek to curtail Soviet fishing activity or with competitors whose collective operations threaten the existence of a particular species. On January 22, 1971, Canada and the Soviet Union concluded two Agreements patterned somewhat after Soviet-American arrangements. The first appertains to cooperation in fisheries in the northeastern Pacific off the Canadian coast;[118] the second lays down provisional rules of navigation and fisheries safety for the same region.[119] The latter Agreement was extended automatically for two more years on April 22, 1973, but the former was subject to review in February 1973. The USSR, Norway, and Iceland signed an Agreement on the Regulation of the Fishing of Atlanto-Scandian Herring[120] at Moscow on February 25, 1972 for the rest of that calendar year.

With respect to Chile, which claims a 200-mile fishing zone, the Soviet Union has begun to collaborate in the development of coastal fisheries pursuant to the 1971 Chilean-Soviet Agreement on Collaboration in the Development of Fisheries.[121] Collaboration on the Soviet side is taking the form of building or converting fishing ports and related commercial installations, carrying out joint investigations of fish stocks, and training Chilean specialists in modern fishery technology. Article 9 of the Fisheries Agreement establishes a Chilean-Soviet Fishing Commission to elaborate and coordinate measures under the treaty. A key feature of the Agreement is the

116. *Id.*
117. *Id.* at 204.
118. Canadian-Soviet Agreement on Cooperation in Fisheries in the Northeastern Pacific Ocean off the Coast of Canada, *done* Jan. 22, 1971.
119. Canadian-Soviet Rules on Provisional Rules of Navigation and Fisheries Safety in the Northeastern Pacific Ocean off the Coast of Canada, *done* Jan. 22, 1971.
120. NEW DIRECTIONS, *supra* note 33, at 449–50.
121. *Done* Sept. 7, 1971, 11 INT'L LEGAL MATERIALS 1 (1972).

Article 5 provision for chartering of Soviet fishing trawlers by a Chilean corporation. Payment for the use of the vessels and for operating expenses is made in fish meal or other ocean products.[122] Rather than complicate relations with Chile by precipitating a confrontation over the 200-mile zone (to which the USSR remains implacably opposed in law), the Soviet Ministry of Fisheries recoups some benefit by chartering its modern, efficient trawlers. The pattern could well be imitated elsewhere.

Recent Soviet proposals on fisheries are disposed to favor the developing countries and coastal species. The operative principle seems to be that developing States should have a preferential right, in the areas adjacent to a twelve-mile territorial sea or fishing zone, to the share of fish stocks they are capable of harvesting. The balance of the stock would be available to distant-water fleets, having due regard for conservation requirements.[123]

Conclusion

The developments considered above in large part confirm the proposition that the Soviet Union has moved closer to the views and policies of the leading Western maritime powers than at any time in her history. Precipitous judgments about patterns of recent events all too often look absurd in retrospect. At the moment it would appear that Soviet attitudes toward the law of the sea are still evolving. These modifications, however, have not occurred simultaneously with Western perceptions of augmented Soviet capabilities at sea, nor have changes within the realms of public and private maritime law taken place either at the same time or in the same degree.

Soviet attitudes toward the international law of fisheries correlate most directly with the physical deployment and problems of the fishery fleet. The years 1958–60 were crucial in this regard, and it is not unreasonable to speculate that a better appreciation of the im-

---

122. The text of the Soviet-Chilean contract and criticism thereof expressed in the Chilean Senate is reproduced in 11 INT'L LEGAL MATERIALS 947–53, 1156–68 (1972).

123. *See* Soviet comments in the Seabed Committee, U.N. Doc. A/AC. 138/SR. 27, at 9–10 (March 27, 1972), and the Draft Articles on fishing submitted July 21, 1972, by the USSR to Sub-Committee II of the Seabed Committee, U.N. Doc. A/AC. 138/SC. II/L. 6 (July 18, 1972). Similar views are set forth in a Declaration on Principles of Rational Exploitation of the Living Resources of the Seas and Oceans in the Common Interests of All Peoples, adopted at a Conference of Ministers of Eastern European States in Moscow on July 6–7, 1972, and circulated to the Seabed Committee as U.N. Doc. A/AC. 138/85 (1972).

plications of a broad territorial sea for high seas fisheries was partly responsible for attenuating the Soviet position at the 1960 Geneva Conference on the Law of the Sea. On the whole, however, 1967–69 seems to have been a decisive period for Soviet policymakers. Legislation on the continental shelf[124] and a new Merchant Shipping Code[125] were enacted, the latter having been debated for many years; initiatives were taken to sound out opinion about a new law of the sea conference;[126] protests against unilateral claims of coastal States were more frequent;[127] the 1954 London Convention concerning oil pollution was ratified;[128] bilateral maritime arrangements with States on a variety of issues were concluded;[129] and several traditional Soviet doctrinal positions showed signs of being reconsidered.[130] And while it would be an exaggeration to say that views of the USSR and the Western maritime powers in the United Nations Seabed Committee are the same, they coincide to a degree that would have been inconceivable fifteen years ago.

There is no reason to believe that the enlargement of Soviet maritime interests is a passing phenomenon. The Ninth Five-Year Plan calls for an increase of 47 percent in edible fish production and a 40 percent increase in sea transport cargo turnover.[131] Offshore mining will probably turn out to be a venture of the 1980s, but Soviet expressions of interest are substantial, and exploratory activities are underway. Naval developments remain enigmatic. Augmented Soviet naval capabilities have so far had a minor impact on traditional Soviet views. The classic juxtaposition of the attitudes of naval powers and those of other coastal nations toward the law of the sea no longer applies to the Soviet Union, which has departed from the latter category, but has not moved firmly into the former category. Soviet attitudes toward the classification of straits would suggest that vulnerability from the sea continues to be of great concern. Plainly, the evolution of a Soviet high seas mentality, if that is to be, is far from complete.

124. Edict on the Continental Shelf of the U.S.S.R., Feb. 6, 1968, [1968] 6 Ved. Verkh. Sov. S.S.S.R. Item 40 (Supreme Soviet USSR).
125. [1968] 39 Ved. Verkh. Sov. S.S.S.R. Item 351 (Supreme Soviet USSR).
126. *See* W. Butler, *supra* note 1, at 45.
127. *Id.*
128. Vodnyi Transport, Sept. 23, 1969, at 3.
129. *See* W. Butler, *supra* note 1, at 236–37.
130. *Id.* at 198–202.
131. On Marine transport, see Kozhevnikov, Ryzhov, & Karasev, *Zadachi morskogo i rechnogo transporta v deviatoi piatiletke,* Planovoe Khoziaistvo, No. 1, at 47–52 (1973).

Canada's Jurisdiction over the Arctic and the Littoral Sea

L. C. Green

Introduction

One of the most persistent issues in international law has been that of the freedom of the seas and, concomitantly, the extent of State jurisdiction over the littoral sea. As long ago as the first half of the seventeenth century, Grotius was upholding the principle of freedom in his *Mare Liberum*,[1] while Selden proclaimed the sovereign right to a closed sea in his *Mare Clausum*.[2] Even though it appeared for a period that Bynkershoek's exposition of the three-mile–cannon-shot rule might be accepted,[3] it was clear that some States were not prepared to accept this rule in connection with their customs and fisheries legislation.[4]

Controversies over the limits to national jurisdiction were responsible for the failures of the League of Nations[5] and the United Nations codification efforts in this field. Even though the latter's Conference was able to adopt a Convention on the Territorial Sea and Contiguous Zone[6] and another on the High Seas,[7] the inability to define the extent of the territorial sea made it impossible to determine

---

1. H. GROTIUS, MARE LIBERUM (1633). *But see* H. GROTIUS, DE JURE BELLI AC PACIS 214 (The Classics of International Law ed. 1925), which states: "It seems clear . . . that sovereignty over a part of the sea is acquired . . . insofar as those who sail . . . along the coast may be constrained from the land . . . ."
2. J. SELDEN, MARE CLAUSUM (1635). *See also* J. BOROUGHS, THE SOVEREIGNTY OF THE BRITISH SEAS (1631).
3. Bynkershoek states: "[T]he control from the land ends where the power of men's weapons ends . . . ." C. VAN BYNKERSHOEK, DE DOMINIO MARIS 44 (The Classics of International Law ed. 1923). *See also* C. VAN BYNKERSHOEK, QUESTIONUM JURIS PUBLICI 54 (The Classics of International Law ed. 1930); Walker, *Territorial Waters: The Cannon Shot Rule*, 22 BRIT. Y.B. INT'L L. 210 (1945).
4. For particular reference to English claims, see, *e.g.*, T. FULTON, THE SOVEREIGNTY OF THE SEA (1911). *See generally*, W. MASTERSON, JURISDICTION IN MARGINAL SEAS WITH SPECIAL REFERENCE TO SMUGGLING (1929).
5. L.N. Doc. C.74.M.39V (1929), 24 AM J. INT'L L. 25, 27–29 (Supp. 1930).
6. Convention on the Territorial Sea and the Contiguous Zone, *done* Apr. 29, 1958, [1964] 2 U.S.T. 1606, T.I.A.S. 5639, 516 U.N.T.S. 205.
7. Convention on the High Seas, *done* Apr. 29, 1958, [1962] 2 U.S.T. 2312, T.I.A.S. 5200, 450 U.N.T.S. 11.

where the high seas began. The Conference on the Law of Sea in 1960 was no more successful.[8] Since then the situation has become even more complicated as individual States seek to expand their territorial and fishery limits[9] to an extent that well-nigh negates any idea of a free ocean. Even before the Law of the Sea Conferences took place, a number of Latin American States had put forward claims to areas as much as 200 miles from their coasts, and coercive measures were resorted to in some cases to enforce such claims.[10] At the time of the 1958 Conference, however, little was heard of such claims, and it might have been thought that they had been abandoned, the United Nations Convention having limited the territorial sea and the contiguous zone to twelve miles. Nevertheless, these claims have now been revived, as evidenced by Ecuador's recent fining of American and Canadian vessels for fishing within 200 miles of its coasts[11] and the claim put forward by Iceland, its rejection by the United Kingdom and the Federal Republic of Germany, and the defiance of Iceland of the interim injunctions delivered by the International Court of Justice.[12]

In addition to the traditional disputes, the question of the limits of national jurisdiction over the sea has become more complicated in recent years because of the threat presented to the ecology, economy, and geography of littoral States by oil spills and the discharge of other deleterious matter into the high seas. In fact, it was concern for the environment, aggravated by the threat of excessively large tankers proceeding through northern waters, that has been largely responsible for the controversy concerning Canada's claim to sovereignty over the Arctic and the exercise of jurisdiction in the name of pollution control at some distance from Canadian coasts.

According to the Canadian Government, what it has done by extending its territorial limits does not amount to an assertion of

---

8. U.N. Conf. on the Law of the Sea, Geneva, 1960, Official Records, U.N. Doc. A/Conf. 19/8, at 9 (1960). For a short summary of this conference, see R. Dhokalia, The Codification of Public International Law 308–11 (1970).

9. For a collection of such claims, see United Nations, National Legislation and Treaties Relating to the Territorial Sea, the Contiguous Zone, the Continental Shelf, the High Seas, and to Fishing and Conservation of the Living Resources of the Sea, U.N. Doc. ST/LEG/SER. B/15 (1970).

10. See, e.g., Note, *Territorial Waters and the Onassis Case,* 11 The World Today 1 (1955).

11. The Times (London), Nov. 17, 1972, at 10, col. 5.

12. Fisheries Jurisdiction Case, [1972] I.C.J. 12, 30, 12 Int'l Legal Materials 290, 300 (1973). For the earlier phases of this dispute, see Green, *The Territorial Sea and the Anglo-Icelandic Dispute* 9 J. Pub. L. 53 (1960). See also Re Red Crusader, 35 I.L.R. 485 (Comm'n of Enquiry, Denmark–United Kingdom 1962).

sovereignty but is merely an exercise of functional protection or jurisdiction, which the Court described in the *Bernadotte Case,*[13] with the
aim of limiting pollution and preserving fisheries, natural resources,
and safety.[14] Canada does not in any way deny the importance of the
freedom of the seas or the need for free access to scientific information and coastal waters.[15] It contends, however, that to assert
these freedoms in the sense of classical international law amounts to
an oversimplification that ignores the significance of commercial
considerations and national security.[16] It has also been pointed out on
behalf of Canada that its acceptance of the tweve-mile limit is more
consistent with current views on the international law of the sea than
is the continuing insistence by the United States on a three-mile
limit.[17] For example, at the 1960 Geneva Conference a wider than
three-mile limit was defeated only by virtue of the mathematical com-
plications involved in securing the necessary majority.[18] If such a
conference took place today, the twelve-mile limit would almost certainly secure overwhelming acceptance.

Historical Background Supporting Canada's Claim over the Arctic

International law still observes the maxim *falsa demonstratio non
nocet.* This means that, even though a State asserts that it is not
claiming sovereignty, if in fact what it claims and concomitantly carries into effect constitutes the essence of sovereign authority, then,
despite the disavowal, sovereignty is in fact in issue. In the case of the
Arctic and Canada's stand in relation thereto, however, it may well be
that Canada did not need to assert a claim to sovereignty, for this
might already have existed. If this is so, it might be argued that by
virtue of the statements made at the time of the enactment of the
Arctic Waters Pollution Prevention Act,[19] the amendment of the

13. Reparation for Injuries Suffered in the Service of the United Nations Case, [1949]
    I.C.J. 174.
14. Prime Minister Trudeau, Press Conference, Apr. 8, 1970, 9 INT'L LEGAL
    MATERIALS 600 (1970).
15. 1 [1969] PARL. DEB., H.C. 39–40 (remarks by Prime Minister Trudeau).
16. Trudeau, *supra* note 14.
17. Canadian Reply to the United States, Apr. 17, 1970, 9 INT'L LEGAL MATERIALS
    607 (1970).
18. A two-thirds majority of those present and voting was required, and the six-plus-six
    proposal failed by only one vote. A clear 12-mile fishing limit was rejected by 39 to
    36, with 13 abstentions. U.N. CONF. ON THE LAW OF THE SEA, GENEVA 1960,
    OFFICIAL RECORDS, U.N. Doc. A/CONF. 19/8, at 151, 165 (1960).
19. REV. STAT. CAN. 1970, c. 2 (1st Supp.).

Canada Shipping Act[20] and the extension of territorial limits by the Territorial Sea and Fishing Zones Amendment Act,[21] Canada in fact cast unnecessary doubts upon its title.[22] Traditionally, sovereignty has depended on discovery, the assertion of claims, territorial contiguity, occupation, and jurisdiction, with the greatest emphasis being placed on the last. In order to ascertain whether Canadian sovereignty exists with regard to the Arctic, it is necessary to examine past history.

The issue of sovereignty acquired topical significance three or four years ago when it became known that United States companies interested in exploiting the Arctic mineral resources had printed maps suggesting that areas Canada had always considered to be part of the Canadian Arctic were under United States administration or *terra nullius* and thus open to sovereignty by whosoever could establish title. Public interest in Canada was stimulated and emotions inflamed when it was learned that Humble Oil intended to send a giant tanker through northern waters, alleging that the waters in question were international and that the intention was to open a year-round channel. Canadian public opinion was so aroused over a threat to Canadian sovereignty and to the ecology that the Government of Canada had no choice but to enact legislative proposals to extend the breadth of the territorial waters and at the same time retreat from the jurisdiction of the International Court of Justice.[23]

The problem of Arctic sovereignty and Canadian title is not new. In 1925 David Hunter Miller said: "[W]hereas Canada makes a precise and definite claim of sovereignty, no other country . . . has announced any claim whatever. Furthermore, the appearance of these islands on the map as a northern extension of the Canadian mainland is a visible sign of an important reality—namely, that many of them are quite inaccessible except from or over some Canadian base. With her claim of sovereignty before the world, Canada is gradually extending her actual rule and occupation over the entire area in question."[24]

Parliamentary Debates

Some fifty years earlier the Canadian Parliament had sought a clear declaration as to the extent of Canada's boundaries. In 1878 in a Joint

20. Rev. Stat. Can. 1970, c. 38 (1st Supp.).
21. Rev. Stat. Can. 1970, c. 45 (1st Supp.).
22. For further discussion, see Green, *Canada and Arctic Sovereignty*, 48 Can. B. Rev. 740 (1970).
23. *Compare* Declaration of Apr. 7, 1970, [1969–1970] I.C.J.Y.B. 55 *with* Declaration of Sept. 20, 1969, [1968–1969] I.C.J.Y.B. 46.
24. Miller, *Political Rights in the Arctic,* 4 Foreign Affairs 47, 51 (1925).

Address to the Queen, the Canadian Government requested that these be limited "[o]n the East by the Atlantic Ocean, which boundary shall extend towards the North by Davis Straits, Baffin's Bay, Smith's Straits and Kennedy Channel, including all the islands in and adjacent thereto. . . . On the North the Boundary shall be so extended as to include the entire continent to the Arctic Ocean, and all the islands in the same westward to the 141 meridian west of Greenwich; and on the North West by the United States Territory of Alaska."[25] This was followed by an Order in Council in 1880 by which "all British possessions on the American continent, not hitherto annexed to any colony" were transferred to Canada.[26] The vagueness of this statement arises from the then indefinite state of knowledge as to what comprised "the entire continent to the Artic Ocean"—islands, water, ice, or a continental landmass. With scientific knowledge as it then was, it was not uncommon for boundaries to be indicated in broad general terms rather than to be clearly demarcated. A somewhat similar instance is to be seen on the Indo-Chinese frontier.[27]

Canadian politicians clearly felt that the Order in Council was not sufficiently specific, and in 1907 Senator Poirier sought "a formal declaration of possession of the lands and islands situated to the north of the Dominion, and extending to the north pole."[28] Although this effort failed, the Poirier statement has come to be regarded as the basis of the sector demarcating Canadian sovereignty extending from the northward reaches of Canada to the Pole. This sector has been escribed as "deceptively simple," consisting of

a base line or arc described along the Arctic Circle through territory unquestionably within the jurisdiction of a temperate zone state, and sides defined by meridians of longitude extending from the North Pole south to the most easterly and westerly points on the Arctic Circle pierced by the state. Under the theory, nations possessing territory extending into the Arctic regions have a rightful claim to all territory—be it land, water or ice—lying to their north. This claim springs from the geographical relationship of the claimant state to the claimed territory; the two areas must be contiguous along the Arctic Circle.[29]

The meridians in question are 141 and 60 west, 141 being the line referred to in the Anglo-Russian Treaty of 1825[30] and the Alaska Pur-

---

25. 1 [1878] PARL. DEB., SEN. 903.
26. REV. STAT. CAN. 1970, Appendices, No. 14.
27. *See, e.g.,* Green, *Legal Aspects of the Sino-Indian Border Dispute,* [1960] CHINA Q., No. 3, at 42, 46–47, 53, 56.
28. [1906–07] PARL. DEB., SEN. 266.
29. Head, *Canadian Claims to Territorial Sovereignty in the Arctic Regions,* 9 McGILL L.J. 200, 202–03 (1963).
30. 12 BR. FOR. STATE PAPERS 38 (1824–25).

chase Agreement of 1867[31] in relation to its prolongation to the Frozen Ocean.

Even at the time of Poirier's statement there was evidence that Canada regarded something more than geographical contiguity as essential for sovereignty, for according to Minister of Trade and Commerce Richard Cartwright, the federal government had sent an expedition, established posts, "exercised various acts of dominion," "levied customs duties and . . . exercised our authority over the various whaling vessels they have come across, which, I think, will be found sufficient to maintain our just acts in that quarter."[32] The instances of jurisdictional activity to which he refers are similar to those recognized as grounding evidence for sovereignty in such cases as those concerning the Palmas [Miangas][33] and the Minquiers and the Ecrehos [Ecrehou] Reefs.[34]

Geographical Titles and Sovereignty

As to contiguity and similar geographic claims, it must be remembered that some of these, like the hinterland theory or South American claims to Antarctica, are nothing but attempts to dress predatory assertions in respectable garb. Of the sector theory *simpliciter,* particularly in its application in Antarctica, Judge (then Professor) Waldock has commented that they are

based fundamentally on the principle of geographical continuity of territory. Indeed they are nothing more or less than new examples of the old hinterland doctrine.

Arctic sectors, although they also are based on the principle of proximity, are really examples of another proximity doctrine, "contiguity". . . . "Contiguity" is the name given to the doctrine sometimes invoked in support of claims to islands lying near to a state's territory but outside its territorial waters. The mere proximity of the island to the claimant state is represented as a geographical connexion between the two lands as a ground for including the island within the sovereignty of the nearby state. . . .

It is not believed that . . . [the] sector doctrine can by itself be a sufficient legal root of title. The hinterland and contiguity doctrines as well as other geographical doctrines were much in vogue in the nineteenth century. They were invoked primarily to mark out areas claimed for future occupation. But,

31. Convention Ceding Alaska, Mar. 30, 1867, 2 W. MALLOY, TREATIES 1521 (1910).
32. [1906–07] PARL. DEB., SEN. 274.
33. The Island of Palmas Case (United States v. the Netherlands), 2 U.N.R.I.A.A. 829, Hague Court Reports (Scott 2d ser.) 83 (Perm. Ct. Arb. 1928).
34. The Minquiers and Ecrehos Case, [1953] I.C.J. 47.

by the end of the century, international law had decisively rejected geographical doctrines as distinct legal roots of title and had made effective occupation the sole test of the establishment of title to new lands. Geographical proximity, together with other geographical considerations, is certainly relevant, but as a fact assisting the determination of the limits of an effective occupation, not as an independent source of title.[35]

Waldock was merely applying to the problem of determining sovereignty over the Falklands and similarly situated territory the principles that Judge Huber had already outlined in the *Palmas Case:*

The principle of contiguity, in regard to islands, may not be out of place when it is a question of allotting them to one State rather than another, either by agreement between the Parties, or by a decision not necessarily based on law; but as a rule establishing *ipso jure* the presumption of sovereignty in favour of a particular State, this principle would be in conflict with what has been said as to territorial sovereignty and as to the necessary relation between the right to exclude other States from a region and the duty to display therein the activities of a State. Nor is this principle of contiguity admissible as a legal method of deciding questions of territorial sovereignty; for it is wholly lacking in precision and would in its application lead to arbitrary results. . . .

There lies, however, at the root of the idea of contiguity one [further] point which must be considered . . . . [I]n the exercise of territorial sovereignty there are necessarily gaps, intermittence in time and discontinuity in space . . . . [Such is the case with territories that are] partly uninhabited or as yet partly unsubdued. The fact that a State cannot prove display of sovereignty as regards such a portion of territory cannot forthwith be interpreted as showing that sovereignty is inexistent. Each case must be appreciated in accordance with the particular circumstances.

. . . [I]nternational arbitral jurisprudence . . . would seem to attribute greater weight to—even isolated—acts of display of sovereignty than to continuity of territory.

. . . [But] we must distinguish between . . . the act of first taking possession, which can hardly extend to every portion of territory, and . . . the display of sovereignty as a continuous and prolonged manifestation which must make itself felt throughout the whole territory.[36]

Huber observed that these manifestations may assume "different forms, according to conditions of time and place. Although continuous in principle, sovereignty cannot in fact be exercised at every moment on every point of a territory. The intermittence and discontinuity compatible with the maintenance of the right necessarily differ," depending upon whether the regions involved are inhabited or uninhabited or enclosed within territories in which sovereignty is in-

35. Waldock, *Disputed Sovereignty in the Falkland Island Dependencies,* 25 BRIT Y.B. INT'L L. 311, 341, 342 (1948).
36. 2 U.N.R.I.A.A., at 854–55, Hague Court Reports (Scott 2d ser.) at 111–12.

contestably displayed or accessible from, for instance, the high seas. "It is true that neighboring States may by convention fix limits to their own sovereignty, even in regions such as the interior of scarcely explored continents where such sovereignty is scarcely manifested, and in this way may each prevent the other from any penetration of its territory."[37]

In the *Eastern Greenland Case* the Permanent Court of International Justice emphasized as important factors

the intention and will to act as sovereign, [and] some exercise or display of such authority . . . [for tribunals have] been satisfied with very little in the way of the actual exercise of sovereign rights, provided that the other State could not make out a superior claim. This is particularly true in the case of claims to sovereignty over areas in thinly populated or unsettled countries. [What should be borne] in mind [is] the absence of any claim to sovereignty by another Power, and the Arctic and inaccessible character of the uncolonized parts of the country.[38]

The *Minquiers and Ecrehos Case* merely serves to emphasize how little in the way of the exercise of jurisdiction will suffice to establish sovereignty when no effective contrary claims can be substantiated. As already indicated, Canada was exercising administrative functions in the Arctic as early as 1927. In the 1925 amendments to the Northwest Territories Act,[39] which controlled entry into the Arctic, the Minister of the Interior stated that Canada claimed all the territory between longitude 60 and 141 "right up to the Arctic."[40] From 1925 on, permits were required to enter the Arctic, the Canadian Minister to the United States having written to Secretary of State Stimson that "this requirement has been fulfilled by the scientists and explorers of many nations since that date."[41] Moreover, some States have expressly recognized Canadian title to some of the islands in the region, for example, Norway with respect to the Sverdrup Islands.[42]

United States Attitude Toward Arctic Claims

Canada is not the only country to have put forward claims to sovereignty in Arctic regions. Imperial Russia and the Soviet Union

---

37. 2 U.N.R.I.A.A. at 840, Hague Court Reports (Scott 2d ser.) at 94.
38. Legal Status of Eastern Greenland Case, [1933] P.C.I.J., ser. A/B, No. 53, at 46, 50, 51.
39. An Act to Amend the Northwest Territories Act, Stat. Can. c. 48 (1925).
40. 4 [1925] Parl. Deb., H.C. 4084.
41. Letter from Minister Phillips to Sec. Stimson, Nov. 21, 1929, in 1 G. Hackworth, Digest of International Law 463 (1940).
42. Norwegian Note of Aug. 8, 1930, *id.* at 465, [1930] Can. T.S. No. 17.

both claimed the epicontinental shelf of the Russian mainland and all that lay between the coast and the Pole as being within Russia's sector.[43] The Soviet claim to Wrangel Island was recognized by Britain and Canada in 1924, although it would seem that the United States, the property of whose nationals had been confiscated by the Russians, has not abandoned its own claim.[44] Equally, the United States has occasionally indicated its unwillingness to recognize Canada's title. In 1924, the Secretary of the Navy stated: "In my opinion, it is highly desirable that if there is in that region land, whether inhabitable or not, it should be the property of the United States. . . . I cannot view with equanimity any territory of that kind being in the hands of another Power."[45] Hardly a legal basis for title, and in 1925 Hunter Miller observed:

The United States has never officially made any claim to any known Arctic is-lands outside of our well recognized territory. . . .

. . . .

As to the islands now known and lying north of the Canadian mainland, the average American would have no objection to the Canadian title. . . . The only other possibilities would be something in the nature of *terra nullius,* an un-satisfactory sort of ownership by everybody, or else ownership by the United States. No public sentiment here would favor either, as against Canada.

. . . .

. . . [T]he probability is that few of the claims thus far made to lands hitherto discovered will be questioned.

. . . .

So while it cannot be asserted that Canada's title to *all* these islands is le-gally perfect under international law we may say that as to almost all of them it is not now questioned and that it seems in a fair way to become complete and admitted.[46]

This would appear to be clear acknowledgment by a senior Depart-ment of State adviser that Canada's title is better than that of any other State, including the United States.

Hunter Miller explained the extent of the Canadian title: "[T]he official Canadian claim, so far as it relates to the unknown, is in the nature of a notice before discovery and before occupation. What Canada says is that if Arctic lands be found—found by anyone—east of 141° and west of 60° and Davis Strait, they are Canadian, or will

43. For the Ukase of 1916, as reissued in 1924, see W. LAKHTINE, RIGHTS OVER THE ARCTIC 43–45, App. (1928); 1 J. DEGRAS, SOVIET DOCUMENTS ON FOREIGN POLICY 476 (1951). *See also* T. TURACOUZIO, SOVIETS IN THE ARCTIC 348 (1938).
44. G. HACKWORTH, *supra* note 41, at 464.
45. G. SMEDAL, ACQUISITION OF SOVEREIGNTY OVER POLAR AREAS 68 (1931).
46. Miller, *supra* note 24, at 54, 52, 54, 53.

be."[47] Here we have almost a forerunner of Article 2 of the Geneva Convention on the Continental Shelf,[48] and the vagueness inherent in that document reflects Miller's view of the Canadian geographical limits, for he says that "the expression 'as far as the Frozen Ocean' [in the 1825 Treaty and the Alaska Purchase] is vague enough . . . to make it at least arguable that the line runs as far as the 141st meridian itself runs, and that means to the North Pole[,]"[49] including presumably land, islands, water, and whatever else may be there. Given the view of the Court in the *Eastern Greenland Case* concerning contrary claims, it is well to remember that, as pointed out by the Secretary of State for External Affairs in the House of Commons in 1959, "since 1900 there is no record of any dispute with Russia or America" concerning the ownership of any portion of the Canadian Arctic.[50]

Though other countries did not put forward positive claims to any part of the Canadian Arctic, Canadian spokesmen were asserting "what we have we hold,"[51] and that according to the generally recognized sector principle "our sovereignty extends right to the Pole within the limits of the sector."[52] Prime Minister St. Laurent declared in 1953 that "we must leave no doubt about our active occupation and exercise of our sovereignty in these lands right up to the Pole."[53] As has been seen, international law is satisfied with relatively little by way of active occupation and exercise of sovereignty. Canada has in fact sent Northwest Mounted Police patrols into the area,[54] established well inside the Arctic Circle government posts that issued licenses (some of which were requested by the United States[55]), and exacted taxes. Moreover,

[d]uring the [Second World] War the United States Government asked permission of Ottawa to establish certain weather and emergency installations in Upper Frobisher Bay and Cumberland Sound on Baffin Island [of which Miller said there could be no challenge to Canada's title], as well as air bases at Coral Harbor on Southampton Island and Cape Dyer on Baffin Island. This permission was, of course, granted, but as a war measure on a temporary

47. *Id.* at 56.
48. Convention on the Continental Shelf, *done* Apr. 29, 1958, [1964] 1 U.S.T. 471, T.I.A.S. No. 5578, 499 U.N.T.S. 311.
49. Miller, *supra* note 24, at 59.
50. 2 [1959] PARL. DEB., H.C. 1822.
51. 2 [1922] PARL. DEB., H.C. 1750.
52. 3 [1938] PARL. DEB., H.C. 3081.
53. 1 [1953] PARL. DEB., H.C. 700.
54. 1 FOREIGN REL. U.S. 571 (1940).
55. G. HACKWORTH, *supra* note 41, at 463.

basis, subject to the right of Canada to replace the stations, and to the stipulation that all permanent facilities with respect to the air bases, having been paid for in full, should become the property of Canada after the war.[56]

The establishment of D.E.W. line stations has not affected the situation, and United States vessels servicing such stations must apply to the Canadian Government for waivers of the Canada Shipping Act[57] before proceeding.[58]

Statements concerning Canada's sovereignty in this area may be multiplied.[59] They show that jurisdiction was intended to extend to the Pole, islands, land, sea, and ice. As Charles Cheney Hyde has pointed out:

It is not apparent why the character of the substance which constitutes the habitual surface above [sea] level or its lack of permanent connection with what is immovable, should necessarily be decisive of the susceptibility of a claim to sovereignty of the area concerned. This should be obvious in situations where the particular area is possessed of a surface sufficiently solid to enable man to pursue his occupations thereon and which also in consequence of its solidity and permanence constitutes in itself a barrier to navigation as it is normally enjoyed in the open sea [as is clearly the case with the pack ice in the area].

States at times endeavor to acquire rights to sovereignty over polar areas by acts which would be regarded as inadequate were the regions sought to be acquired within the temperate zone. . . . [A]t the present time an aspirant to sovereignty over a polar region . . . may, by means of aircraft and a variety of other devices, make its will felt throughout a district which it claims as its own, and by such process establish its supremacy therein. . . . Canada is understood to approve generally of the sector system. . . . The Dominion appears, however, to deem it necessary to fortify its position by other processes, and to endeavor in fact to exert a degree of administrative control over adjacent polar areas which it claims as its own. . . . [M]eans of communication and transportation as well as control are such [today] as to justify a demand for more than an assertion of dominion by a mere symbolic act, and to cause the perfecting of a right of sovereignty to be dependent upon the exercise of some measure of control over the area involved. . . . In the Arctic regions it must be acknowledged that the sovereign of a contiguous area of land that projects itself well into the Arctic Circle is in a relatively advantageous position to make its supremacy felt within or over an extensive yet unoccupied area. That potentiality which is attributable in large part to geographical considerations, strengthens the applicability of the sector principle in the

56. Pearson, *Canada Looks "Down North,"* 24 FOREIGN AFFAIRS 638, 641 (1946). Miller has asserted that there was no question as to Canada's title to Upper Frobisher Bay and Cumberland Sound. Miller, *supra* note 24, at 51.

57. Canada Shipping Act, REV. STAT. CAN. c. S9, 1970.

58. 3 [1957] PARL. DEB., H.C. 3186. (remarks by Prime Minister St. Laurent).

59. *See, e.g.,* Green, *supra* note 22, at 746–52.

North Polar regions. Yet it points also to the conditions to be met in order to preserve if not perfect a right of sovereignty therein, as, for example, by Canada or Russia.[60]

Hyde recognized that attempts might be made to extend the application of the Monroe Doctrine to the area, but reminded his readers

that as the United States has not sought to interfere under cover of that doctrine with the acquisition of rights of sovereignty over areas that were not at the time deemed to be capable of settlement by such peoples [from temperate climes], it has left the problem pertaining to the polar regions untouched. . . . The strength of [this contention] will be weakened if the polar regions prove to be susceptible to control by means that fall short of occupation or settlement, and if such control is sought to be exercised by a non-American power. . . . The extension of Canadian assertions of dominion to adjacent polar areas however wide, if deemed to satisfy the normal requirements for the acquisition of rights of sovereignty over polar areas, may not be regarded by the United States as infringing upon the operation of the Monroe Doctrine. . . . [T]he northward strides of Canada may not, therefore, be looked upon as those of a non-American power.[61]

Moreover, it should be noted that while the Monroe Doctrine may be a well-established postulate of United States policy, it has not received acknowledgment as a recognized principle of international law.

Recent Assertions of Sovereignty

It is clear that there have been sufficient statements made on Canada's behalf to demonstrate her intention to assert sovereignty. Furthermore, Canada has indulged in enough administrative actions to substantiate that claim by jurisdictional acts that have received recognition in accordance with accepted international standards. Until recently, there had been adequate acknowledgment by other States to indicate that Canada's Arctic sovereignty was accepted. Assertions brought forward in recent years on the basis of new political aims or developing economic interests would be rejected in any international arbitration. In addition, Canada has also successfully exercised judicial functions within the area. One of the most significant instances of this was the decision of the Northwest Terri-

60. 1 C. HYDE, INTERNATIONAL LAW CHIEFLY AS INTERPRETED AND APPLIED BY THE UNITED STATES 348, 350, 354, 355, (2d ed. 1945).
61. *Id.* at 290, 291, 290.

tories Territorial Court in *Regina v. Tootalik*[62] in November 1969, when Mr. Justice Morrow applied the Northwest Territories Game Ordinance of 1960 to "the sea-ice offshore from Pasley Bay" in waters frequently icebound even in summer. He pointed out that "it is not declarations of sovereignty that count so much as the actual day-by-day display of sovereign rights." The court referred to the patrols of the Northwest Mounted Police well into the Arctic area[63] and earlier judicial decisions relating to offenses committed on both land and sea ice hundreds of miles within the area as evidence of this daily assertion of sovereignty.[64]

The recent exercise of jurisdiction by a United States court over an individual accused of murder within what has been habitually regarded as the Canadian—or perhaps Danish—sphere of jurisdiction does not affect the principle here involved.[65] In the first place, should there have been a usurpation of Canadian rights, the failure of the Canadian authorities to protest would not in any way affect the Canadian right to have made such a protest, a point already made in the *Palmas Case*.[66] Further, many States assert, and international law does not deny, the right to exercise jurisdiction over nationals for selected offenses, including murder, bigamy, and the like,[67] when committed outside the national territorial jurisdiction, provided they are able to subject the offender to their actual authority or the national jurisdiction authorizes trials in absentia. Further, the fact that the accused has been wrongly arrested does not in any way affect the right of the court to exercise its jurisdiction, or of the State whose sovereignty has been infringed to protest should it so desire.[68] As has been indicated, the Canadian judiciary has not been willing to treat the sea ice as high seas, nor has W. Lakhtine, one of the leading commentators on Arctic sovereignty. In 1930 he wrote: "the doctrine of the high seas, if applied to the Arctic Ocean, is quite unsatisfactory. Sovereignty should attach to the Polar States over the Arctic Ocean within their sectors of

---

62. 71 W.W.R. (n.s.) 435 (1970).
63. 71 W.W.R. (n.s.) at 439.
64. For some of the earlier decisions, see J. SISSONS, JUDGE OF THE FAR NORTH (1968).
65. N.Y. Times, May 11, 1971, at 24, col. 5; Globe & Mail (Toronto), May 11, 12, 19, 1972, at 11, 29, 8 respectively.
66. 2 U.N.R.I.A.A. at 843.
67. *See, e.g.,* Andersen v. United States, 170 U.S. 481 (1898), 1 J. MOORE, DIGEST OF INTERNATIONAL LAW 932 (1906), 2 A. McNAIR, INTERNATIONAL LAW OPINIONS 183 (1956); Rex v. Earl Russell, [1901] A.C. 446.
68. *See, e.g.,* Kerr v. Illinois, 119 U.S. 436 (1886); *Ex Parte* Elliott, [1949] 1 All E.R. 373 (K.B.); Eichmann v. Atty. Gen. of Israel, 36 I.L.R. 277 (Supreme Court, Israel 1962).

attraction. The jurisdiction, however, should be qualified by the assurance to foreign Powers of the right of innocent passage of all naval vessels, although the littoral State should have the right to regulate, control and even prohibit hunting and fishing. . . ."[69]

Modern developments have raised the issue of ecological damage and danger to the littoral State itself, an issue that surely goes far beyond the risk of injury to livestock suitable for hunting and fishing. If protection of the wildlife may justify control or even prohibition of foreign activities, then surely this is equally true of measures to prevent pollution. The *Anglo-Norwegian Fisheries Case,*[70] together with Article 4 of the Geneva Convention on the Territorial Sea, has indicated that where a sinuous coast exists, straight base lines may be used to indicate the measuring line for the territorial sea. True, it would be contrary to the Convention if such lines were drawn so as to cut off another State from the high seas, but it would be stretching geographical facts unrealistically to suggest that, even if the entire area were a closed Canadian water, it in fact separated United States territorial seas from the high seas. It is true that Lakhtine talked of innocent passage, but, as the *Corfu Channel Case* indicated, when a natural waterway joins two parts of the high seas, it must be used for international navigation to be deemed an international area.[71] Until the first passage of the *Manhattan* there was no suggestion that such an international passage existed through Canada's Arctic waters. In any case, innocent passage pertains only to the territorial sea, not to internal or national waters. The islands to the north of Canada's mainland constitute an archipelago whose headlands may be joined to determine base lines. If this is done, the waters are national and not territorial, and navigation would then clearly depend on Canadian consent. But even if the waters are territorial, Canada would still be entitled to require all shipping passing through to observe the local regulations concerning peace, good order, and security of the littoral areas. These regulations would clearly include antipollution regulations and other measures concerning the preservation of natural resources or the local ecology.[72] Even the Spitzbergen Treaty, which confirmed alien rights in the area placed under Norwegian sovereignty, subjected such rights "to the observance of local laws and

---

69. Lakhtine, *Rights over the Arctic*, 24 AM J. INT'L. L. 703, 713 (1930).
70. Anglo-Norwegian Fisheries Case, [1951] I.C.J. 116, 129 *as corrected by* Erratum of Oct. 22, 1956.
71. The Corfu Channel Case, [1949] I.C.J. 4, 28. *See also* Convention on the Territorial Sea and the Contiguous Zone, art. 16 (4).
72. See Green, *International Law and Canada's Anti-Pollution Legislation*, 50 ORE. L. REV. 462 (1971).

regulations."[73] Therefore, it is submitted that any recent Canadian act or statement purporting to be an assertion of sovereignty is, in fact, nothing but an exercise of an existing sovereignty.

Canadian Legislation to Protect the Ecology

The issue of Canadian rights and sovereignty in the Arctic as well as that of Canada's right to control access to her coasts came to the forefront when Prime Minister Pierre Trudeau announced in October 1969 that there was a need for legislation to protect the ecological balance in the Canadian Arctic. Trudeau indicated that this would be done by antipollution regulations, accompanied by an extension of Canada's territorial sea to twelve miles and the establishment of new fisheries zones. He declared that the proposed measures were intended as an expression of Canada's regard for the ecological balance existing in the Arctic archipelago:

This legislation we regard, and invite the world to regard, as a contribution to the long-term and sustained development of resources for economic and social progress.

We also invite the international community to join with us and support our initiative for a new concept, an international and legal régime designed to ensure to human beings the right to live in a wholesome natural environment. ... A combination of an international régime, and the exercise by the Canadian government of its own authority in the Canadian Arctic, will go some considerable distance to ensure that irreparable harm will not occur as a result of negligent or intentional conduct.

Canadian activities in the northern reaches of the continent have been farflung and pronounced for many years, to the exclusion of the activities of any other government.[74]

In 1929 Canada accepted the Optional Clause of the Statute of the Permanent Court of International Justice for all disputes other than those relating to Canadian domestic matters as determined by international law, inter-Commonwealth disputes, and issues for which the parties had chosen other means of settlement.[75] Canada has never appeared before the Permanent Court or the International Court of Justice but has always advocated international judicial settlement as a means for deciding issues. Prime Minister Trudeau himself stated:

73. Treaty concerning the Archipelago of Spitzbergen, Feb. 9, 1920, art. 3, 2 L.N.T.S. 8.
74. 1 [1969] PARL. DEB., H.C. 39.
75. *See* note 23 *supra*.

Membership in a community, . . . imposes . . . certain limitations on the activities of all members. For this reason, while not lowering our guard or abandoning our proper interests, Canada must not appear to live by double standards. We cannot, at the same time that we are urging other countries to adhere to régimes designed for the orderly conduct of international activities, pursue policies inconsistent with that order simply because to do so in a given instance appears to be to our brief advantage. Law, be it municipal or international, is composed of restraints. If wisely construed they contribute to the freedoms and the well being of individuals and of States. Neither States nor individuals should feel free to pick and choose, to accept or reject, the laws that may for the moment be attractive to them.[76]

Despite this statement, Trudeau expressed dissatisfaction with the present state of international law in connection with pollution, the Arctic, and the like. He observed that international law had not kept pace with the advance of technology and the growing threat of pollution and that Canada was prepared to help develop the law by taking measures of its own.[77]

Doubtful or developing law often prompts parties in dispute to refer their problem for solution to a judicial tribunal, and it is certainly uncommon—and unknown on the international level—for a tribunal to deny itself jurisdiction because there is no law concerning the issue or there is a lacuna that it is unable or unwilling to fill.[78] Nevertheless, in accordance with Trudeau's statement, Canada added to its acceptance of the jurisdiction of the International Court of Justice a reservation concerning "disputes arising out of or concerning jurisdiction or rights claimed or exercised by Canada in respect of the conservation, management or exploitation of the living resources of the sea, or in respect of the prevention or control of pollution or contamination of the marine environment in areas adjacent to the coast of Canada."[79] Unfortunately, this exclusion of jurisdiction was completely unnecessary. After all, at the time it was made, the only potential opponents to the Canadian policy were the Soviet Union and the United States. Insofar as the Soviet Union is concerned, it has not accepted the jurisdiction of the Court, and present indications are that it is unlikely to do so. As for the United States, since the Statute of the Court specifically refers to jurisdiction based on reciprocity, and the Court itself in the *Norwegian Loans Case*[80] upheld the validity of the

76. 1 [1969] Parl. Deb., H.C. 38–39.
77. Trudeau, *supra* note 14, at 601–04.
78. *See, e.g.,* Naulilaa Case (Germany v. Portugal), 2 U.N.R.I.A.A. 1011, 1016 (1928); Green, *Filling Lacunae in the Law,* 29 Malayan L.J. xxviii (1963).
79. Declaration of Apr. 7, 1970, [1969–70] I.C.J.Y.B. 55.
80. Case of Certain Norwegian Loans, [1957] I.C.J. 9.

domestic jurisdiction reservation based on self-interpretation, any suit brought by the United States could have been effectively halted by Canada's reciprocal invocation of the United States reservation based on the Connally Amendment.[81] While it might not have been politically wise for Canada to appear to block the jurisdiction of the Court on purely formal grounds, this probably would have appeared better than seeking a procedure that denied it any role and left Canada appearing to adopt a policy of eclecticism, deciding unilaterally which rules of law it found desirable and which it rejected or felt were to be interpreted as not supporting the Canadian stance. This is the more unfortunate in view of the Prime Minister's own statements.

Extension of the Territorial Sea to Twelve Miles

The Canadian proclamation extending territorial limits to the twelve-mile mark is, as Trudeau indicated, fully consistent with present trends in international law and is probably more in line with what is now accepted in State practice than is the policy of those States that still appear wedded to the three-mile line. Thus, the Canadian Government observed that "in 1958 . . . some 14 States claimed a 12-mile territorial sea, whereas by 1970 some 45 States have established a 12-mile territorial sea and 57 States have established a territorial sea of 12 miles or more. Indeed, the three-mile territorial sea is now claimed by only 24 countries. . . ."[82] It is perhaps unfortunate that the proclamation concerning the width of the territorial sea was made part of the package with the antipollution legislation and the modification in the Declaration concerning the jurisdiction of the Court. In view of the trend in international law and the general acceptance of the twelve-mile limit, Canada could have announced such an extension of its territorial sea, including within its purview the Canadian Arctic as constituting territory within its jurisdiction. This might well have resulted in less furor than followed the Government's action.

Right To Enforce Antipollution Legislation

It is now generally accepted that States may extend their jurisdiction over visiting vessels to ensure that they do not endanger coastal in-

81. [1971–1972] I.C.J.Y.B. 84
82. Summary of Canadian Note of Apr. 16, 1970, 9 INT'L LEGAL MATERIALS 607, 609 (1970).

terests. What Canada has done, however, is to assert her right to introduce and enforce antipollution legislation throughout Arctic waters extending 100 miles off the Canadian coasts. The extent to which international law tolerates the purported exercise of State jurisdiction over foreign vessels on the high seas is somewhat limited. The Canadian legislation was indeed something of an innovation, but probably no more so than the issuance by the United States of the Truman Proclamation in 1945.[83] In fact, since Canada set the precedent, a number of States have introduced similar protective measures[84] and multilateral negotiations have taken place that are fully in line with the Canadian statements regarding the need to preserve the interests of mankind and calling for international cooperation.[85] Just as the Truman Proclamation set in motion a trend that was acknowledged in the *Abu Dhabi Arbitration*[86] and formally accepted as a part of international law by the Geneva Convention on the Continental Shelf, so it would appear that the Canadian legislation has served as a somewhat similar trailblazer.

Ecology and Pollution Control

Ecology was put forward as one of the grounds to justify the antipollution measures. Perhaps all that need be said here is that a State may enact regulations to establish shipping safety control zones and law down standards to be complied with by any vessel entering such zones.[87] In some places this may constitute an apparent limitation upon the freedom of the seas insofar as foreign ships are concerned. However, if States are permitted to declare areas of the high seas closed to foreign shipping in order to prevent fall-out danger from nu-

---

83. Presidential Proclamation No. 2667, Policy of the United States with Respect to the Natural Resources of the Subsoil and Sea Bed of the Continental Shelf, Sept. 28, 1945, 3 C.F.R. 67 (1943–48 Comp.), *reprinted in* 40 Am. J. Int'l L. 45 (Supp. 1946).

84. Among the most recent measures is that by Great Britain applying pollution control regulations 50 miles around its coasts. *See* The Times (London), Dec. 16, 1972, at 3, col. 1.

85. *See, e.g.,* Brussels Convention on Establishment of an International Fund for Compensation for Oil Pollution Damage, 11 Int'l Legal Materials 284 (1972); Report of the Stockholm Environmental Conference, U.N. Doc. A/CONF. 48/14 (1972); London Convention on Dumping Wastes at Sea, 11 Int'l Legal Materials 1294 (1972).

86. Petroleum Development Ltd. v. Sheikh of Abu Dhabi, [1951] 18 I.L.R. 144 (Lord Asquith of Bishopstone, Umpire).

87. For further discussion, see Green, *supra* note 72.

clear testing or to ensure proper security during naval exercises, there appears to be no reason why a State cannot impose safety limitations upon vessels entering areas of the sea where their activities, if not so controlled, might endanger the health or economic security of the coastal State.

One aspect of the Canadian legislation might, however, appear to infringe traditional rules of international law. The regulations are to apply even to vessels "owned or operated by a sovereign power other than Canada," unless there are satisfactory grounds to believe that "appropriate measures have been taken by or under the authority of that sovereign power to ensure the compliance of such ship with, or with standards substantially equivalent to[,]" those established within the control zones "and that in all other respects all reasonable precautions have been or will be taken to reduce the danger of any deposit of waste resulting from the navigation of such a ship" in such a zone.[88] But this provision is not as radical as it may seem, for a State is responsible in nuisance, and possibly also in negligence, if it allows damage to be caused to another State as a result of an activity within its jurisdiction, as was made clear in the decisions of the *Trail Smelter Arbitration*.[89] A vessel owned by a foreign sovereign is clearly within the jurisdiction of that sovereign, even though it may be on the high seas. This is no less true when the vessel is passing through another State's national waters. A passage that is likely to cause damage can hardly be described as innocent if the ship is aware of this potential and refuses to comply with safety regulations.

Where the general application of antipollution measures is concerned, it must be remembered that to allow a State to apply its protective measures only after an accident of the kind involved in an oil spill is to permit action when it is almost certain to be too late. For such measures to be effective they must be allowed on an anticipatory basis. The Geneva Convention on the Territorial Sea in Article 24 (i) (a) allows a State to introduce within its contiguous zone, limited to twelve miles from the base line, control measures only for customs, fiscal, immigration, or sanitary purposes. In addition, Article 6 of the Conservation Convention recognized the coastal State's "interest in the maintenance of the productivity of the living resources in any area of the high seas adjacent to its territorial sea."[90] This allows a coastal

---

88. Pollution Prevention Act, Rev. Stat. Can. 1970, c. 2 (1st Supp.), 9 Int'l Legal Materials 543, 547–48 (1970).

89. Trail Smelter Arbitration (United States v. Canada), 3 U.N.R.I.A.A. 1905 (1938); *id.* at 1938 (1941).

90. Convention on Fishing and Conservation of the Living Resources of the High Seas, done Apr. 29, 1958, [1966] 1 U.S.T. 138, T.I.A.S. 5969, 559 U.N.T.S. 285.

State to introduce conservation measures limiting the quantity of fish that may be taken from the high seas adjacent to the coastal sea. Apparently, while a State is allowed to take measures to conserve the fishery, it is not permitted to take the steps necessary to keep the fishery alive—unless the term *sanitary* is broadly defined. Normally this term is understood in international law as referring only to the preservation of health and does not include things like sewage pollution. But since a people's health may be affected by pollution of the fishery in the contiguous zone or by polluted waters lapping its shores, there seems to be no reason why a State should not be able to take preventive measures within its contiguous zone, and, if it may do so within the contiguous zone, it may clearly do so within its territorial sea.

The U.N. Charter and Self-Defense to Prevent Pollution

There remains the question of preventive self-defense in its economic context. Before the establishment of the United Nations, anticipatory self-defense was clearly recognized in international law, as is evidenced by many of the reservations to the Kellogg-Briand Pact.[91] The statement made in the United States Senate Report dealing with the Pact's ratification reaffirms the concept of self-defense in that "each nation is free at all times and regardless of the treaty provision to defend itself, and is the sole judge of what constitutes self-defense and the necessity and extent of the same."[92]

It is often contended that Article 51 of the Charter of the United Nations removed the right of anticipatory self-defense, since it refers to the "inherent right of . . . self-defense if an armed attack occurs." But even those who argue that this is an exhaustive statement are prepared to concede that, in view of the existence of long-range missiles ready for use, "the difference between attack and imminent attack may now be negligible."[93] Moreover, it is difficult to believe that the draftsmen of the Charter, all practical politicians, intended to impose a limit upon the freedom of action of their own States insofar as self-defense measures are concerned, while acknowledging that nonmembers, who, despite Article 2 (6), may ignore the Charter completely, remained under no restriction.

The Charter acknowledges that the right of self-defense is inherent.

91. Aug. 27, 1928, 4 W. MALLOY, TREATIES 5130 (1938), 94 L.N.T.S. 57.
92. J. WHEELER-BENNETT, DOCUMENTS ON INTERNATIONAL AFFAIRS 6 (1928).
93. I. BROWNLIE, INTERNATIONAL LAW AND THE USE OF FORCE BY STATES 368 (1963).

It therefore exists by customary law and pertains to every State by reason of its statehood. Since the United Nations guarantees to every member its continued existence and independence, it is most unlikely that it would impose upon members any obligation that might militate against their continued ability to exist. The draftsmen were aware of such incidents as the Italian attack upon Ethiopia, and it is unlikely that they favored a regulation obliging a State to await an attack that might be conclusive before granting to that State any right to defend itself. Logic and legal reasoning reject any idea that membership in the United Nations would place a member at a disadvantage as compared with nonmembers or even in its relations with other members. Since the Charter does not define self-defense but merely describes it as an inherent right, there is much to be said for accepting the view held in 1928, that a potential victim might indeed strike the first blow against a potential aggressor, even though this might present an appearance of his being the aggressor.[94] Such an approach would be in accord with Daniel Webster's comments at the time of the *Caroline* incident,[95] the holding of the Court in the *Nuremberg Judgment*,[96] and the views of such writers as Julius Stone,[97] D. W. Bowett,[98] and Myres McDougal and Florentino Feliciano.[99]

It is submitted that despite the wording of Article 51, the Charter does not in fact reduce a State's right to take preventive action against a threat to its peaceful existence. The fact that the United Nations seems, for political and ideological reasons, to have departed from the classical interpretation does not alter the situation. In order to determine whether a State's inherent right to self-defense has been limited by the Charter, one should examine the intention of the participants at the time the Charter was drafted. Further, since Article 51 refers to the functions of the Security Council, it does not really matter what the General Assembly says on this issue.

The above argument is not limited merely to the preservation of the physical existence of a State. As Bowett has pointed out, "in practice the interest which a state may have in the safe preservation of the national economy, of its essential economic interests, may be equally as

---

94. *See* Green, *Armed Conflict, War and Self-Defence,* 6 Archiv des Völkerrechts 387 (1957).
95. 2 J. Moore, Digest of International Law 412 (1906).
96. International Military Tribunal (Nuremberg), Judgment and Sentences, Oct. 1, 1946, 41 Am J. Int'l. L. 172, 207 (1947).
97. J. Stone, Aggression and World Order 44 (1958).
98. D. Bowett, Self-Defence in International Law 234 (1958).
99. M. McDougal & F. Feliciano, Law and Minimum World Public Order 235, 237 (1961).

great as the interest in safeguarding its territory, its political independence, or its people."[100] This view is to be found in Article 15 of the Bogotá Charter, which expressly prohibits any "form of interference or attempted threat against the personality of the State or against its political, economic and cultural elements."[101] Moreover, the concept of economic aggression is constantly referred to in the councils of the United Nations[102] and has appeared in the definition of aggression proposed by Bolivia: "unilateral action whereby a State is deprived of economic resources derived from the proper conduct of international trade or its basic economy is endangered so that its security is affected."[103] It would seem, therefore, that there is no reason why self-defense should not be invoked against threats to economic interests.

In its opinion in the *Austro-German Customs Union Case*[104] the Permanent Court emphasized that economic sovereignty was a part of State sovereignty and that a threat against economic independence was in fact a threat against State independence.[105] Six of the eight judges constituting the majority expressly stated that "since [the proposed regime] would be calculated to threaten the independence of Austria in the economic sphere, [it] would constitute an act capable of endangering the independence of that country."[106]

Conclusion

If the World Court's view is a true reflection of the relationship of economic independence to independence as such, so that a threat to the one is a threat to the other—and there is no reason to doubt this contention—it should follow that if a State may take anticipatory preventive measures of self-defense to preserve its political or territorial independence, it may equally take similar measures with regard to potential threats to its economic independence and, since economic

100. D. Bowett, *supra* note 98, at 106.
101. Organization of American States Charter, art. 15, [1951] 2 U.S.T. 2394, T.I.A.S. 2361, 119 U.N.T.S. 3.
102. N.Y. Times, May 6, 1958, at 17, col. 2 (Bolivia against the Soviet Union); *id.*, Oct. 2, 1958, at 29, col. 1; *id.*, July 19, 1964, at 1, col. 4 (Cuba against the United States); *id.*, Dec. 5, 1972, at 1, col. 4 (Chile against the United States).
103. Report of the Secretary General on the Question of Defining Aggression, U.N. Doc. A/2211, at 52 (Oct. 3, 1952). *See also* Report of the 1956 Special Committee on the Question of Defining Aggression, 12 U.N. GAOR Supp. 16, at 30, U.N. Doc A/3574 (1957), for the Soviet view.
104. [1931] P.C.I.J., ser. A/B, No. 41.
105. *Id.* at 52.
106. *Id.* at 53. *See also id.* at 73 (Separate Opinion of Judge Anzilotti).

survival depends upon economic well-being, against any action that constitutes a substantial threat to its economic welfare. As a result, a State still enjoys the right under customary international law to take action even before a maritime incident has occurred. It seems to be generally accepted, notably in recent treaties,[107] that a coastal State may even sink a foreign ship on the high seas when a maritime incident threatens pollution. This being so, it hardly seems worthwhile affirming that such coastal State may prevent an accident by turning away a foreign ship from its coastal waters, or impose safety standards as preconditions for its entry. It would appear that Canadian actions relating to the Arctic and concerning antipollution measures with respect to the waters off the Canadian coasts are in accord with existing rules of international law, either on the ground that such action was consistent with Canada's existing rights as a sovereign exercising jurisdiction over its own territory or because such measures were within the limits laid down by customary law for the defense of legitimate State interests.

107. *See, e.g.,* Brussels Convention relating to Intervention on the High Seas in Cases of Oil Pollution Casualties, Nov. 29, 1969, arts. 1(1), 3(d), (e), 9 INT'L LEGAL MATERIALS 25 (1970).

Contributors

William E. Butler is Reader in Comparative Law at the University of London, and sometime Senior Scholar at the Moscow State University.

Northcutt Ely is Attorney at Law, Washington, D.C.

Luke W. Finlay is Attorney at Law, New York, N.Y.

L. F. E. Goldie is Professor of Law and Director of the International Legal Studies Program at Syracuse University.

L. C. Green is University Professor at the University of Alberta.

Robert D. Hodgson is the Geographer of the United States Department of State.

J. Michel Marcoux is Attorney at Law, Washington, D.C.

Terry V. McIntyre is Assistant Geographer of the United States Department of State.

Editors

George T. Yates III is Jervey Fellow in Foreign Law at the Columbia University School of Law.

John Hardin Young is Attorney at Law, Washington, D.C.

Index